Y0-AAV-890

Wiley Series of Practical Construction Guides

M. D. MORRIS, P. E., EDITOR

Jacob Feld
CONSTRUCTION FAILURE

William G. Rapp
CONSTRUCTION OF STRUCTURAL STEEL BUILDING FRAMES

John Philip Cook
CONSTRUCTION SEALANTS AND ADHESIVES

Ben C. Gerwick, Jr.
CONSTRUCTION OF PRESTRESSED CONCRETE STRUCTURES

S. Peter Volpe
CONSTRUCTION MANAGEMENT PRACTICE

Robert Crimmins, Reuben Samuels, and Bernard Monahan
CONSTRUCTION ROCK WORK GUIDE

B. Austin Barry
CONSTRUCTION MEASUREMENTS

D. A. Day
CONSTRUCTION EQUIPMENT GUIDE

Harold J. Rosen
CONSTRUCTION SPECIFICATION WRITING

Gordon A. Fletcher and Vernon A. Smoots
CONSTRUCTION GUIDE FOR SOILS AND FOUNDATIONS

Don A. Halperin
CONSTRUCTION FUNDING: WHERE THE MONEY COMES FROM

Walter Podolny, Jr. and John B. Scalzi
CONSTRUCTION AND DESIGN OF CABLE-STAYED BRIDGES

John P. Cook
COMPOSITE CONSTRUCTION METHODS

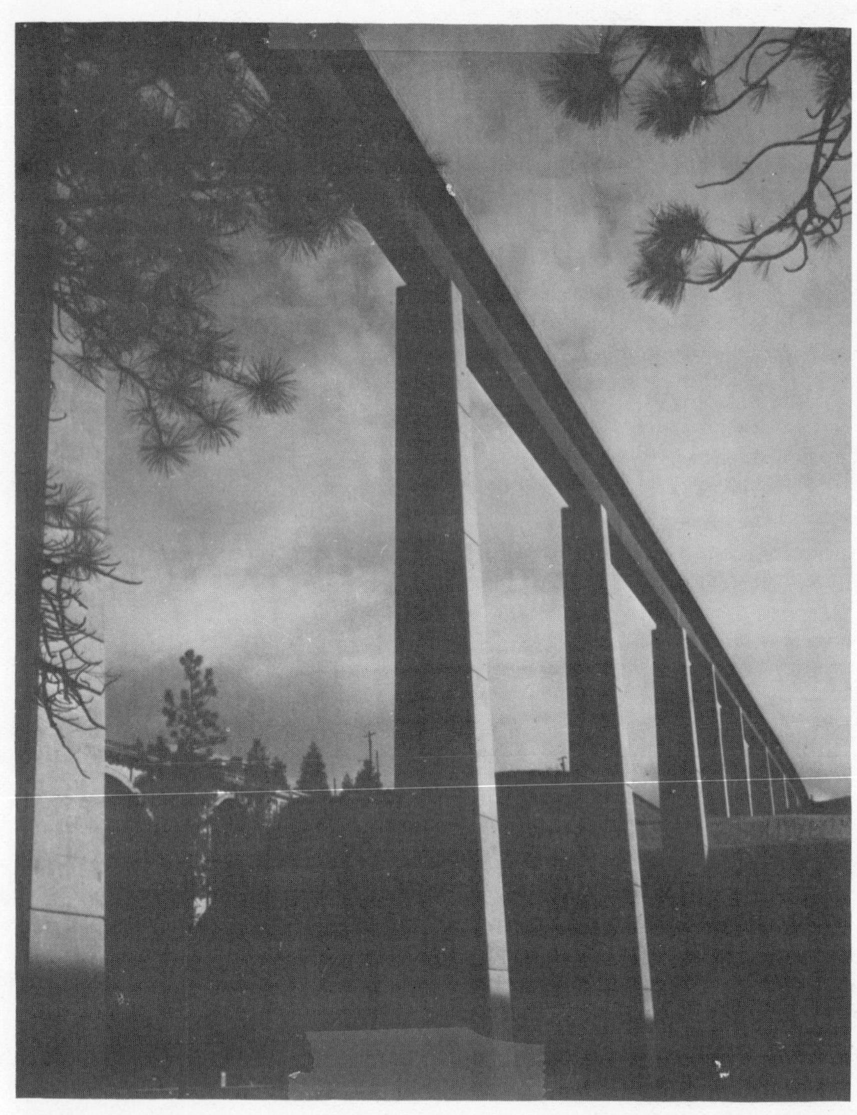

Latah Creek Bridge—composite box girder. Courtesy of Howard, Needles, Tammen, and Bergendoff.

COMPOSITE CONSTRUCTION METHODS

JOHN P. COOK, P.E.
Jacob Lichter Professor of
Engineering Construction
University of Cincinnati

A Wiley-Interscience Publication

JOHN WILEY & SONS
New York London Sydney Toronto

Library of Congress Cataloging in Publication Data:

Cook, John Philip.
Composite construction methods.

(Wiley series of practical construction guides)
"A Wiley-Interscience publication."
Includes bibliographies and index.
1. Composite construction. I. Title.
TA664.C65 624'.18 76-26020
ISBN 0-471-16905-6

Printed in the United States of America

10 9 8 7 6 5 4 3 2 1

Series Preface

The construction industry in the United States and other advanced nations continues to grow at a phenomenal rate. In the United States alone construction in the near future will exceed ninety billion dollars a year. With the population explosion and continued demand for new building of all kinds, the need will be for more professional practitioners.

In the past, before science and technology seriously affected the concepts, approaches, methods, and financing of structures, most practitioners developed their know-how by direct experience in the field. Now that the construction industry has become more complex there is a clear need for a more professional approach to new tools for learning and practice.

This series is intended to provide the construction practitioner with up-to-date guides which cover theory, design, and practice to help him approach his problems with more confidence. These books should be useful to all people working in construction: engineers, architects, specification experts, materials and equipment manufacturers, project superintendents, and all who contribute to the construction or engineering firm's success.

Although these books will offer a fuller explanation of the practical problems which face the construction industry, they will also serve the professional educator and student.

M.D. Morris, P.E.

Preface

Today, with the decline in the supply of high-grade raw materials, rising prices, and an increasingly tight, skilled labor market, architects and engineers are turning with increasing frequency to the various forms of composite construction to gain strength, economy, and beauty.

The ground work for the acceptance of composite construction was laid by the painstaking research and development of men such as Newmark, Pavlo, Siess, Voellmy, Viest, Fisher, Winter, Slutter, and many others whose contributions are found in the technical literature and are repeated here as references. Without their work, there would be no need for this book.

This is not intended solely as a design book. Complete design procedures are found in the references noted and in the various codes. The object here is to treat design and construction as "full brothers" in completing a structure. There are some design examples included, but they are kept as simple as possible with a minimum of confusing subscripts.

I extend my thanks to the American Institute of Steel Construction (AISC), the American Concrete Institute (ACI), the American Association of State Highway and Transportation Officials (AASHTO), and the American Institute of Timber Construction (AIIC) for permission to reproduce those portions of their specifications which deal with composite construction. These reproductions, however, are only excerpts and are included for ready reference. A complete copy of the specifications is a necessity in order to handle all aspects of a completed structure. Addresses for the appropriate specification writing bodies are included in the appendix.

Thanks are due to those engineers, architects, and trade associations who furnished illustrations and information for the text. They are specifically acknowledged throughout the book.

Thanks are due also to Lieutenant Kenneth Hover, U.S. Army Corps of Engineers, for permission to quote freely from his excellent thesis on forming and falsework, and to James Simmons, who prepared most of the drawings for this text.

Last and probably most important, my thanks go to Mr. Dan Morris, the Series Editor, for his constant help and prodding.

JOHN P. COOK

Cincinnati, Ohio
December 1976

Contents

COMPOSITE CONSTRUCTION METHODS

1

History and Introduction

1.1 Introduction

Composite construction consists of using two materials together in one structural unit and using each material to its best advantage. A concrete slab connected to a steel beam forms a composite beam. Actually, every reinforced concrete beam is a composite member. The concrete, which is good in compression, takes the compressive forces. The steel, generally placed toward the tension face of the beam, takes all the tensile forces. A concrete-filled pipe column forms a good composite member. The concrete and steel work together to help each other in preventing buckling and supporting the load. Trusses can also be composite members. The number of combinations is almost endless: steel and concrete, timber and concrete, precast and cast-in-place concrete, timber and steel.

Now, newer materials and designs are being used: fiber-reinforced plastics (FRP); various plastics combined with wood chips to form flakeboard or hardboard; plastics combined with colorful aggregates to form decorative panels, stair treads and sink tops.

1.2 Advantages of Composite Construction

Until quite recently, when the term "composite construction" was mentioned, the listener automatically thought of a steel beam and a concrete slab, tied together with a shear connector. Today, however, the designers are branching out to use other types of composites, such as the cellular steel decks, timber and concrete, and composite columns.

Although other forms of composite construction are growing rapidly, the steel-concrete composite beam still has the lion's share of the composite market, and economic comparisons are most easily made using this combination.

In steel-concrete construction, estimates from 15 to 40% savings have been made depending on the design and the method of construction. In terms of steel weight, a quick check can be made for typical buildings by using the ratio of the section modulus of the steel beam to the section modulus of the composite beam from the AISC composite beam tables. Beam weight, of course, is not the total picture. Lighter beams can mean smaller columns and less steel tonnage to be supported, which results in a savings in the cost of foundations. Smaller beams also mean a savings in building height.

Smaller, shallower beams can result in a considerable savings in the cost of a multistory building. In a building with a perimeter of 400 ft and 10 stories high, a savings in depth of only 3 in. in the floor system means a savings of 100 ft^2 of exterior cladding material at each story height. This is a savings of 1000 ft^2 for the entire building. The shallower beams also mean a savings of 3 in. of length of every vertical pipe run, conduit, duct, and soil stack at each story height.

In addition to these obvious advantages, there are other benefits which have been largely overlooked. Since the composite beam provides an increase in the stiffness of the beam, the designer can make a reasonable estimate of this increased stiffness. By planning this increased stiffness into the analysis of the structure, the designer might change the entire structural layout of the building and save an additional 10 to 15% of steel weight.

In steel-concrete composite construction, the concrete slab is tightly connected to the steel beam so that it acts as a diaphragm in helping to resist lateral loads on the structure.

Composite construction does bring with it some added costs. Shear connectors must be bought and installed. In the metal deck composite floors, the concrete is no longer merely a fill material. The concrete is a structural component, and its quality must be controlled.

Composite construction has not been popular with construction men in the past, largely because of the shear connectors. Spiral connectors, which were formerly used, were structurally effective, but were a headache for the contractor. The handling, spacing, and tack welding of the spirals were tricky operations at the job site. Even the channel connectors, which are still accepted by both the AISC and AASHTO specifications, are not easy to install properly.

The modern stud gun has finally tipped the balance in the contractor's favor. The studs are light and easy to handle, and a competent crew can install 600 studs per day for an average cost of about 60 cents per stud. The operation is clean and simple, and it can be done just about as easily in the field as in the fabricating shop.

Other factors must also be considered. Much of construction cost is almost fixed cost. The crane has to be on the job to lift the beam, whether the beam weighs 60 or 80 ppf. Also, because of the lighter beams, construction loads and deflections must be monitored carefully.

For simple span beams, the greatest weight savings are achieved by using unsymmetrical beams with a larger tension flange. Unshored, coverplated beams result in substantial weight savings, but the cost of welding on the coverplates must be considered.

The cost figures shown below are for an office building in the 7 to 15 story range. The comparison is made for one bay of a multibay building. The building is a steel frame with bolted construction for the connections. Coverplates, when used, are welded. The bay size of the building is 32 × 32 ft, with 4 beams per bay. A 4-in. reinforced concrete slab uses conventional formwork. The floor system carries normal office building live loads. The noncomposite member used as a standard for comparison is a W 18 × 64 rolled section. The costs used are 1974 figures.

Beam No. 1

W 18 × 64; noncomposite
Steel weight: 4.1 tons
Structural steel cost: $2337
No studs; no coverplates

TOTAL STEEL COST: $2337

Beam No. 2

W 18 × 50; full composite action
Steel weight: 3.2 tons
Structural steel cost: $1824
Cost of studs (installed): $115
No coverplates

TOTAL STEEL COST: $1939
SAVINGS OVER THE NONCOMPOSITE MEMBER: 17%

Beam No. 3

W 16×26 with 1×4 coverplate; full composite action
Coverplate is used in the center 20 ft of the span
Steel weight: 2.21 tons
Structural steel cost: $1259
Cost of studs (installed): $115
Welding cost for coverplates: $398

TOTAL STEEL COST: $1772
SAVINGS OVER THE NONCOMPOSITE MEMBER: 24%

These two composite beams were designed for full composite action. It will be shown later that the use of partial composite action results in fewer studs per beam, which means more savings.

The reduction in steel tonnage of the beams means smaller loads to the columns and to the foundations. In this particular building, the ratio of live load to dead load is relatively high, so that column sizes would be affected only very slightly.

A similar comparison can be made for timber-concrete composite beams. In the case of a simply supported bridge with a span of 30 ft carrying AASHTO H-20 Loading, the size of the laminated bridge stringer reduces from $6\frac{3}{4} \times 40\frac{1}{2}$ to $6\frac{3}{4} \times 34\frac{1}{2}$ with composite action. This is a savings of about $95 per stringer. This, however, is only the cost of the wood. In order to maintain the advantage of this type of composite construction, the lag bolt shear connectors must be installed in the plant. The connectors themselves are cheap enough, but the labor cost of drilling the holes and installing the bolts in the field quickly nullifies the savings in timber cost.

1.3 Historical Development

In the very broad sense, composite construction began with the early civilizations. The early Assyrians are credited with the use of man's first manufactured building material. They made mud bricks that were reinforced with straw and were probably our first composite members.

Later, the Greeks and Romans used faced or veneered walls in which two materials were combined to use the best properties of each.

However, composite construction as we know it began in the middle of the nineteenth century. In 1840, a patent was issued to William Howe for a composite truss of wood and wrought iron. Four years later a patent was issued to Thomas and Caleb Pratt, also for a composite truss of wood and wrought iron. The difference between the Howe and Pratt trusses was in the configuration. The Howe truss used the iron tension rods as vertical members, and the Pratt truss used the diagonals as the iron tension members. It is a tribute to the soundness of the design that many of these old bridges are still in service as covered bridges on rural roads throughout this country (Fig. 1.1).

Also in the mid-nineteenth century, concrete began to be used, especially in Britain, as fireproofing for iron structural members. These encased beams were the first real composite beams.

Today's engineers like to think of hybrid construction as a modern development. Hybrid construction today is the use of different strength steels in different parts of a structure. Actually, the hybrid concept, which is composite construction, is also well over a century old. In 1840, Squire Whipple, who also provided us with the first complete correct analysis of a jointed truss, is credited with the construction of a hybrid bowstring truss. This structure used a combination of cast iron for compression members and wrought iron for the tension members.

About that same time, Earl Trumbull completed a bridge over the Erie Canal at Frankford, New York. This was a combination truss and suspension system, using cast iron and wrought iron. A few years later Fink and Bollman built composite bridges using cast iron with wrought iron or wood with wrought iron. The Fink structures were not true trusses, but they were actually trussed beams.

The composite beam began with the use of fireproofing systems for floors. In the early part of the twentieth century, several studies in different parts of the world spurred the development of the composite beam. In Canada, beams encased in concrete were investigated by the Dominion Bridge Company in 1923. The National Physical Laboratory in Britain also was conducting tests on encased beams at the same time. The results were published by Scott in 1925. In the United States, a patent was issued to J. Kahn in 1926 for a composite beam. Shortly afterwards, R. A. Caughey published the results of his work on compo-

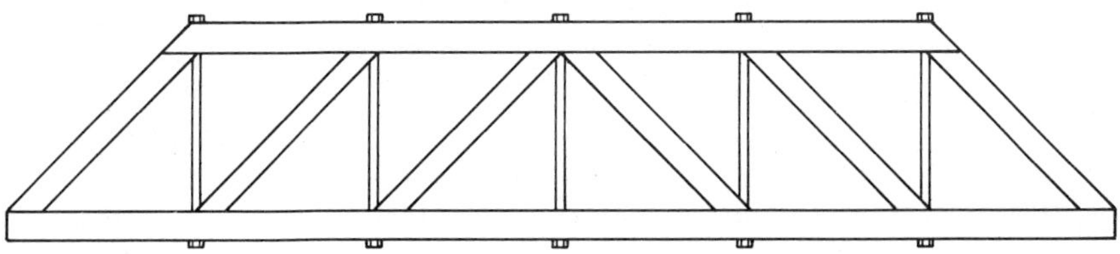

FIGURE 1.1 Howe truss.

site beams of concrete and structural steel in the Proceedings of the Iowa Engineering Society. In 1929, Caughey and Scott collaborated on a paper that dealt with the design of a steel beam and concrete slab. They pointed out the need for a mechanical shear connector to carry the horizontal shear. Their work also included a discussion of both shored and unshored construction.

By the early 1930s, composite construction had become known almost worldwide. About this time the Swiss began using composite construction with spiral shear connectors. In Australia, Knight published a paper on composite bridge design, which included a design of the shear connectors and also a discussion on construction methods. In the United States, the first timber-concrete composite bridges were built during this period.

It is always a big boost for any new construction method when a major specification writing body incorporates the new construction method into its latest code revision. For composite construction, this breakthrough came in two steps. In 1944, the American Association of State Highway Officials included composite construction into its specifications. Then in 1952, the American Institute of Steel Construction included provisions for composite construction for buildings into its code.

After this rapid burst of growth, it was natural that the development of composite construction should slow down. For several years, a great deal of research was conducted into such important, but peripheral, questions as slab uplift, efficiency of different types of shear connectors, natural bond between steel and concrete, vibration and ultimate strength studies, and torsion in composite beams. During this period, most composite construction was of steel and concrete.

In very recent years, architects and engineers have been faced with ever rising prices. Designers are being forced to use their resources of materials and skilled labor more carefully. Consequently, composite construction is just now entering its second great period of expansion and growth.

1.4 Construction-Shoring-Falsework

Throughout the various codes and specifications reference is made to shored and unshored construction. In various writings, savings in steel weight in steel-concrete construction are often shown in terms of shored or unshored construction. In our context, shoring means support under the beams so that the dead load of the slab, formwork, and wet concrete is carried by the temporary supports. The supports should be wedged under the beam so that the beam member does not deflect under the dead load.

Unshored construction goes up faster. After the slab for a given floor level has been cast, only a few days are necessary before the slab is hard enough to be used by other trades, and forming can begin for the next floor level.

In shored construction, the temporary supports must remain in place until the concrete has gained 75% of its design strength, which usually means about 14 days. After the slab is a week or 10 days old, forming and shoring can be done for the next level, but the slab for the next level should not be cast until the composite beam and slab at the first level is capable of supporting the dead load of the second floor. It is generally not good practice to rely on the temporary supports below the first floor to carry the dead load of two stories. Because of the amount of planning necessary, the construction delays, and the cost of shoring, most designers would prefer to use an unshored design, if possible.

The sequence of pictures in Figures 1.2 through 1.6 shows the progress of a shored composite structure. The sequence shows the Sander Hall Residence of the University of Cincinnati. Figure 1.2 shows the

FIGURE 1.2 Sander Hall, University of Cincinnati. (Courtesy of Department of Physical Plant—University of Cincinnati.)

completed facility. The composite structure is the low dining hall building in the foreground. Progress can be estimated by noting the amount of construction completed on the taller building. In this case, construction time was not a big factor. It was only necessary to finish both buildings by the same completion date. Figure 1.3 shows the steel framing almost completed for the building. Shear connectors are already installed on the steel beams. Figure 1.4 shows the slab forms in place for the first floor slab. Figure 1.5 shows the slab completed at the first floor level and formwork being placed for the second floor slab. Figure 1.6 shows the second floor slab complete and the shoring and formwork being placed for the roof slab. In this building, the slab forms were hung from the steel beams, and the shoring consisted of one row of temporary supports, placed one under each beam at midspan. Strictly speaking, this is not fully shored construction. However, since bending moment varies as the square of span length, cutting the dead load span in half for this relatively short span reduces the dead load moment equivalent to almost full shoring.

For slab forming, the trend is away from wood formwork which requires extensive field cutting. Designers are planning beam spacings that use full size plywood sheets and are using more "stay-in-place" metal forms and cellular steel decks.

In bridge work, shoring is seldom feasible. More often than not, there

FIGURE 1.3 Steel framing almost complete. (Courtesy of Department of Physical Plant—University of Cincinnati.)

FIGURE 1.4 (Courtesy of Department of Physical Plant—University of Cincinnati.)

simply is no solid base on which to rest the temporary supports. Figure 3.22 shows a formwork system for a bridge deck slab. This is an effective system, but the weight of the formwork adds another 100 ppf to the dead load on the beam. Compared to the slab and girder weight, this extra load is not too much, but it should be considered. Two other points are worth considering with this type of formwork. The 2 × 12 walers do not provide any lateral support for the compression flange of the beam. Also the exterior beam, depending on how the curb and sidewalk detail is done, may be subjected to torsion because it has the slab on one side only. If the sidewalk is formed and proportioned as shown in Figure 1.7, there will be some compensating dead load on the outside of the exterior beam which will reduce the torsional effect of the slab on the beam.

Shored construction means only that the beam be fully supported until the slab has cured. *Thus,* the beam can be fully supported by resting on the ground and the slab cast. The T beam can then be hoisted into position. Whether this is practical depends on the size of the beam, the capacity of the lifting equipment, and a suitable location for casting the beam.

FIGURE 1.5 (Courtesy of Department of Physical Plant—University of Cincinnati.)

FIGURE 1.6 (Courtesy of Department of Physical Plant—University of Cincinnati.)

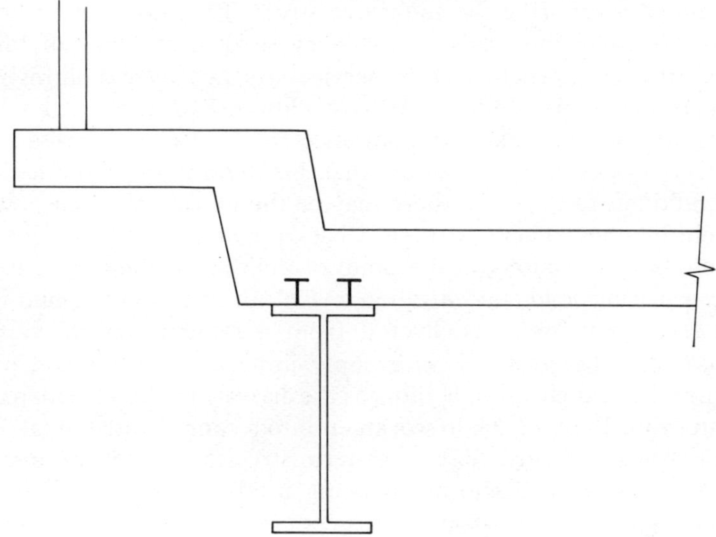

FIGURE 1.7 Exterior bridge stringer.

Prestressing of the steel members has been shown to produce the greatest economy of material. This method is much more valuable in Europe than in the United States. In this country, the extra field labor costs would probably cancel out any savings in material costs. Knowles (1.2) shows several ways that this prestressing can be accomplished. The object of prestressing is to put the bottom flange of the steel section into compression and the top flange into tension. This can be done by jacking the beam upward at midspan before casting the slab or by prestressing cables located below the neutral axis of the steel beam.

1.4.1 Construction Loadings

It is very difficult to generalize on the subject of construction loadings because these loads are different for every project and even differ for identical beams on the same job. After the steel is erected and the slab formwork is in place, construction materials, such as bundles of reinforcing bars, can be placed anywhere on the span. Even two crane operators placing the same load in the same spot may cause different impact loads on the formwork, depending on their ability to lower the load gently. Even the dead load of the concrete slab may vary from beam to beam, depending on the manner of placement.

In the design of the composite member, if the finished member under live load were underdesigned by only a few percent, the designer might be justified in ignoring the slight overstress. This, however, is not the case for construction loads. A relatively small percentage of failures occurs after the structure is in service. By far, most failures occur under construction conditions. A 50 psf allowance for construction loads, while fairly heavy, makes no allowance for load concentrations. The construction-wise designer realizes that this 50 psf is only a simulation of job conditions. On the job, there may be the moving loads of powered concrete buggies. There is often a back-up of several workmen with concrete buggies waiting at the point of delivery of the concrete. This can be a sizeable load concentration which can occur at any point in the span. If the concrete is delivered from a hopper, several yards of concrete may be dropped with some impact on some part of the structure. Even such a simple thing as the arrival of the coffee man at the job can bring 10 or 12 200-lb workmen into a concentrated area. These variable loads can cause high short-term stresses in the beam member, or in the falsework if shoring is being used. Consequently, the wise designer does not gloss over construction stresses as just a temporary condition. They should be treated as probably the most serious design condition.

In the same vein, the contractor should realize that composite design methods are being used more widely now than in the past. The composite member is stiffer and stronger, but it is not a composite beam until the slab has been cast and cured.

Composite construction realizes its economy from the use of lighter, smaller beams which means that composite construction is sensitive to construction loadings. Construction materials and equipment, such as bundles of steel bars, small generators, and compressors, should, if possible, be placed on the span where they will cause the minimum of deflection and vibration.

There are several ways to transport the concrete from the mixer to its final position:

chute
crane and bucket
manual or powered buggies
conveyor or pump

For concrete deposited directly from the chute, it is not impossible to unload as much as 10 yd^3 of concrete in less than 5 min. A rapid rate of chute placement can mean quite a "pileup" of concrete, depending on slump, which can be a significant concentrated load.

Concrete buckets usually hold from 1 to 2 yd of concrete which is placed in a vertical drop at a faster unloading rate than the chute. This can cause a significant impact load and can occur at almost any point in the span.

The motorized buggies used for the delivery of concrete not only cause vertical loads, but also exert horizontal forces on the beam or falsework due to starting and stopping. If buggies are used, the contractor should choose the routes and provide runways so that the loads are distributed safely. If possible, starts and stops for the buggies should be planned parallel to the supporting beams rather than transversely, which would put an out-of-plane load on the compression flange of the beam which is usually unbraced laterally.

One of the most important moving loads in bridge construction is the loading of bridge deck concrete machinery. This consists of a deck finishing machine and perhaps also a deck conveyor. These machines travel the length of the bridge as they place and finish the concrete. They travel along preset rails which may be directly over the exterior girder or on the deck overhang. Figure 1.8 shows a rail with its support fastened to the exterior girder of a large composite bridge. Figure 1.9 shows a close-up of part of the finishing machine for the same bridge.

FIGURE 1.8 Rail for finishing machine. (Coutesy of Structures Subdivision—NYS DOT.)

FIGURE 1.9 Finishing machine. (Courtesy of Structures Subdivision—NYS DOT.)

The machinery manufacturers supply the basic wheel load information for the system, but it is usually up to the contractor's project engineer to fit the system to his particular bridge. In important work, the contractor's project engineer may even have to plot deflection influence lines to insure the proper line and grade in the finished bridge.

References

1.1. Caughey, R. A., "Composite Beams of Concrete and Structural Steel," Proc. 41st Annual Meeting, Iowa Engineering Society, 1929.

1.2. Knowles, P. R., *Composite Steel and Concrete Construction,* Halsted Press, Wiley, New York–Toronto, 1973.

1.3. Hamilton, S. B., "A Short History of the Structural Fire Protection in Buildings, Particularly in England," National Building Studies Special Report No. 27, HMSO, London, 1958.

1.4. *Alpha Composite Construction Engineering Handbook,* Porete Manufacturing Company, North Arlington, N.J., 1949.

1.5. Knight, A. W., "The Design and Construction of Composite Slab and Girder Bridges," *J. Instn. Engrs. Australia,* Vol. 6, No. 1, 1934.

1.6. Scott, W. B., "The Strength of Steel Joists Embedded in Concrete," *Structural Engineer*, No. 26, London, 1925.

1.7. Caughey, R. A., and Scott, W. B., "A Practical Method for the Design of I Beams Haunched in Concrete," *Structural Engineer*, Vol. 7, No. 8, London, 1929.

1.8. Paxson, G. S., "Loading Tests on Steel Deck Plate Girder Bridge with Integral Concrete Floor," Oregon Highway Department, Technical Bulletin 3, 1934.

1.9. Gillespie, P., Mackay, H. M., and Leluau, C., "Report on the Strength of I Beams Haunched in Concrete," *Eng. Journal*, Vol. 6, No. 8, Montreal, 1923.

1.10. Siess, C. P., Viest, I. M., and Newmark, N. M., "Studies of Slab and Beam Highway Bridges, Part III, Small Scale Tests of Shear Connectors and Composite T-Beams," University of Illinois, Eng. Exp. Station Bulletin No. 396, 1952.

1.11. Kinney, J. S., *Indeterminate Structural Analysis*, Addison-Wesley, Reading, Mass., 1956.

1.12. Viest, I. M., Fountain, R. S., and Singleton, R. C., *Composite Construction in Steel and Concrete*, McGraw-Hill, New York, 1958.

1.13. Slutter, R. G., "Composite Steel-concrete Members", in *Structural Steel Design*, Tall, L., Ed., Ronald Press, New York, 1964.

2

How the Composite Beam Works

The composite beam is actually a built-up member, using two different materials and using each material to its best advantage. The usual composite beam consists of three main components: the structural beam (of whatever material), the plate or slab which is added, and some type of connector to hold the beam and slab together.

The structural beam is usually a material which carries tensile stresses efficiently, and the slab is concrete, which has good compressive strength.

Composite beam research has not only proved the efficiency of this type of construction, but also has resulted in greatly simplified design aids, such as the composite beam tables in the current AISC manual.

Construction of the composite beam is no more difficult than ordinary beam and girder construction, but the contractor should know a few of the basic principles of composite design and construction in order to take full advantage of the method.

2.1 Steel-Concrete Composite Construction

In the case of the standard steel-concrete composite member, the units of the composite beam are

1. The steel beam.
2. The concrete slab which acts as a big coverplate.
3. The shear connector.

In ordinary beam and slab construction, the slab is usually designed as a one-way slab, spanning in the transverse direction from beam to beam. In the composite beam, the connectors tie the slab to the beam and force the slab to act in the longitudinal direction. The slab is still designed as a one-way slab and the additional stresses in the longitudinal direction are not considered as affecting the slab design.

In ordinary beam and slab construction, the beam cross section is usually symmetrical, and consequently the neutral axis (NA) is at the middepth of the steel beam. This places the top of the simple span beam in compression and the bottom of the beam in tension. The major effect of the composite action is to force the beam and slab to act together, which shifts the neutral axis of the section upward toward the slab. This leaves the concrete coverplate (the slab) in compression and forces almost the whole steel beam into tension. Thus each of these materials is doing what it does best.

In some cases the neutral axis shifts upward far enough so that it actually lies in the slab. This is an acceptable design, but not the best because part of the slab is in tension and does not contribute to the strength of the section. The case of the neutral axis in the slab can be quite satisfactory in building design where loads are largely static. In bridge construction, if part of the slab is in tension, the continual variation in live load as trucks go over the bridge works the tension cracks in the bottom of the slab open and closed. This can be damaging to the slab. Very often, bridge girders are haunched as shown in Figure 2.2. The haunch design places the main body of the slab some distance above the beam, so that the neutral axis is not likely to fall in the slab.

In the case of a relatively small beam and large slab, in which the neutral axis falls in the slab, the addition of a lower coverplate increases

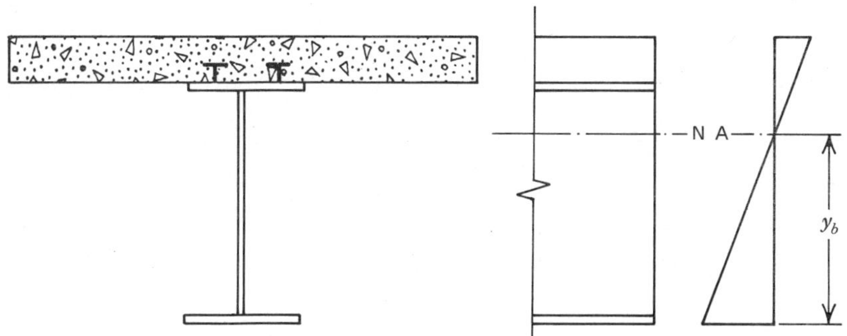

FIGURE 2.1 Steel-concrete composite beam.

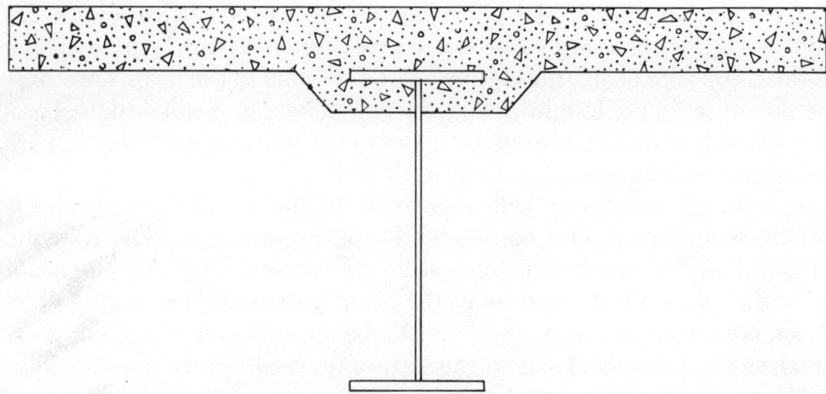

FIGURE 2.2 Haunched girder flange.

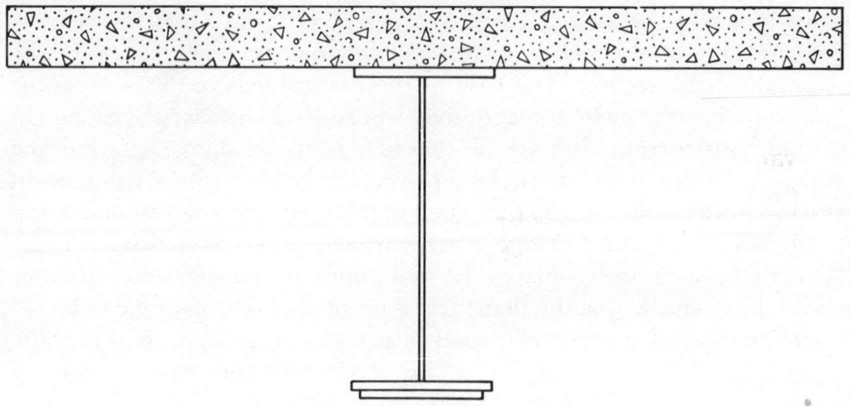

FIGURE 2.3 Coverplated section.

the moment of inertia considerably and shifts the neutral axis downward. The use of coverplated sections is discussed more fully in Chapter 3.

2.1.1 AISC Code Requirements

The actual stresses in steel-concrete composite beams are affected by the method of construction. If the beam is shored, both dead and live loads are carried by the composite section:

$$f_s = \frac{M_D + M_L}{S_{tr}} \tag{2.1}$$

where f_s = steel stress;
M_D, M_L = dead and live load bending moments;
S_{tr} = transformed section modulus of the composite section, with y_b to the bottom fiber.

In the early days of composite construction, the design for unshored construction was simply accomplished by having the steel section carry the dead load and the composite section carry the live load according to

$$f_s = \frac{M_D}{S_s} + \frac{M_L}{S_{tr}} \tag{2.2}$$

According to the current AISC specifications, composite members are designed to carry both dead and live loads, even if the beams are not shored. However, the construction load stresses on the steel beam must not exceed allowable values, and the composite member must meet the requirements of AISC Formula 1.11-2.

$$S_{tr} = \left(1.35 + 0.35\frac{M_L}{M_D}\right) S_s \tag{2.3}$$

If the transformed section modulus computed from this formula exceeds the transformed section modulus supplied, then temporary shoring is not required.

A quick check of a dozen or so sections which might be considered as typical composite beams and with live load to dead load ratios varying from 1:1 to 3:1 shows that few sections fail to meet this requirement. This check shows us that this code requirement, while it limits the ratio of S_{tr} to S_s, is not too restrictive for practical design purposes.

2.2 Analysis of Composite Sections

The analysis of composite sections and the proportioning of members varies slightly with the materials used. In timber-concrete construction, elastic analysis using the transformed section method is used. In steel-concrete construction, elastic analysis is used, but it is modified somewhat by the requirement of AISC Formula 1.11-2. In concrete-concrete construction, either load factor (ultimate strength) or service load (working stress) design can be used.

In the elastic analysis of the composite beam the following assumptions are made:

1. The beam and slab materials are both elastic.
2. These two materials are related by the modular ratio, n.

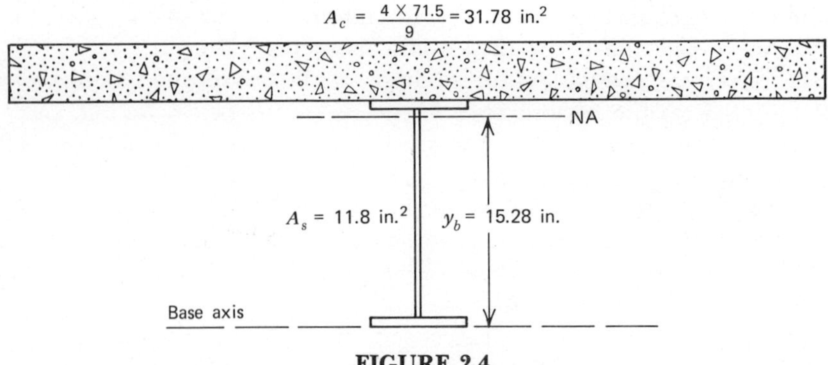

FIGURE 2.4

3. The shear connection provides full interaction between beam and slab.
4. Any concrete below the neutral axis is considered ineffective. (This is reversed, of course, in negative bending regions.)

Referring to Figure 2.4, the usual steps in an elastic analysis are as follows:

1. Take moments of areas about the base of the section to determine location of the neutral axis. Using the modular ratio, transform the slab into an equivalent area of beam material.
2. Calculate the moment of inertia of this transformed section.
3. Calculate the stresses in the slab and beam material under the various loading conditions.

These calculations are illustrated in later chapters for different combinations of materials.

2.3 The Shear Connection

Experience and considerable research have shown that in the steel-concrete composite beam, the natural bond between beam and slab is not strong enough to force the beam and slab to act as a composite member. We know that some natural bond is present, but how effective it is remains unresolved. In the concrete-concrete composite beam, there is a definite value assigned to this bond. This will be discussed later.

In the case of the steel beam, if the beam is fully encased in concrete, it may be designed as a composite member without additional shear connectors, because there is enough bonding surface between steel and concrete to insure that the two materials act together.

In steel-concrete construction under the AISC specifications, shear connectors are no longer designed by the elastic method using VQ/I. Although the beam is proportioned on the basis of elastic analysis, the shear connectors are designed on the basis of ultimate strength. Enough shear connectors should be provided to resist the compressive flange force at the point of maximum moment and to distribute this force to the shear connectors, equally spaced, down to the point of zero moment.

The forces shown in Figure 2.5 are those that exist in the beam at ultimate load.

$$\text{Force } T = A_s F_y$$

$$\text{Force } C_1 = 0.85 f'_c bt$$

Using a factor of 2 to bring these values down to service load levels, the smaller of the two forces is used for the design of the connectors.

$$V_h = \frac{A_s F_y}{2}$$

or

$$V_h = \frac{0.85 f'_c bt}{2}$$

The number of connectors is represented by $n = V_h/q$, where V_h is the lesser of the two values above and q is the tabulated capacity of the connector, determined by testing.

Figure 2.5 shows two different stress distributions at ultimate load, corresponding to different positions of the neutral axis. The beam

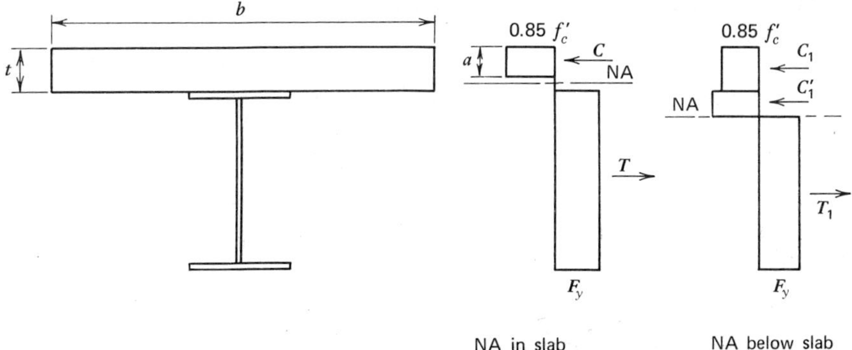

Figure 2.5 Forces in beam and slab at ultimate.

section is proportioned so that the location of the neutral axis in the plastic range is unknown. However, we do not have to locate it. It is much easier to use the code requirement, do the simple calculation for V_h for both cases, and use the smaller value. This can be simply shown from Figure 2.3 as follows:

If the neutral axis is in the slab, from simple equilibrium

$$T = C, C < C_1 \qquad \text{since} \quad C = 0.85 f'_c ba \qquad \text{and} \quad a < t$$

Consequently, $T < C_1$ and is used for design.

If the neutral axis is in the beam,

$$T_1 = C_1 + C'$$

$$C_1 = T_1 - C'$$

$$\therefore C_1 < T_1$$

but

$$T_1 < F$$

Therefore, in both cases we use the smaller of the two values of V_h for design.

Under the AASHTO specifications, shear connectors are designed for fatigue and checked for ultimate strength. In the fatigue design, the familiar formula, $v = VQ/I$, is used for the range of stress under the range of shear due to live load and impact. At any section of the beam, the range of live load shear is taken as the difference in the maximum and minimum shear envelopes. The allowable range of shear on the different types of connectors varies with the number of cycles. These requirements are shown in detail in the Appendix.

The number of connectors provided for fatigue should be checked to insure that enough connectors are provided to develop ultimate strength. The number of connectors for the strength check is as follows:

$$N = \frac{P}{\phi S_u}$$

where N = number of connectors;
$\phi = 0.85$;
S_u = capacity of one connector;
$P = A_s F_y$ or $0.85 f'_c bt$.

Note that these are the same as the AISC expressions except that the factor of 2 is eliminated.

In the concrete-concrete composite member, the shear connection is usually designed by the ultimate strength method, even if the service load method is used to proportion the beam. The shear connection is designed by $v = V/\phi bd$. Although this formula comes from the latest code and is backed up by current research, old-timers will recognize it as similar to the old shear design formula $v = V/jbd$. The ϕ value used in the current code is 0.85, and the j value generally used in the old shear formula was 0.875.

The shear connection in timber-concrete members is designed by the elastic method.

2.4 Effective Flange Width

Virtually every specifying authority has set limits on the width of slab, which may be considered as effective flange width. Actually these effective width limitations are a reasonable simulation of the true effective flange. The shear connectors restrain the slab immediately above the beam so that there is a nonuniform longitudinal stress distribution across the section. Slab length and thickness, load, strain, and Poisson's ratio also affect the true effective width so that it varies at different sections along the beam. Theoretical solutions have been made for the true effective width, but these expressions are complicated and not suited to practical design work. Consequently, the code bodies have set up fairly simple requirements based on span length, slab thickness, and beam spacing. These requirements are shown in detail in later chapters for the different types of members.

2.5 Creep and Stress Relaxation

Creep is the variation of strain with time under constant load or stress. A great deal of research has been done on the creep of concrete members. A generalized creep curve is shown in Figure 2.6. Upon application of the load, there is an instantaneous deformation of the concrete, followed by a time-dependent additional strain.

Stress relaxation is the decay or relaxation of stress under a constant deformation. A typical stress relaxation curve is shown in Figure 2.7.

In composite construction, neither creep nor stress relaxation, in the pure sense, occurs. What really happens is a complex interaction of these two phenomena. To some extent, what happens in the beam depends on the method of construction. If the member is shored during construction, there are no stresses on the bare steel beam. Both dead and live load stresses are carried by the composite section. Under dead load and live loads of long duration, the concrete will creep. In the steel-concrete

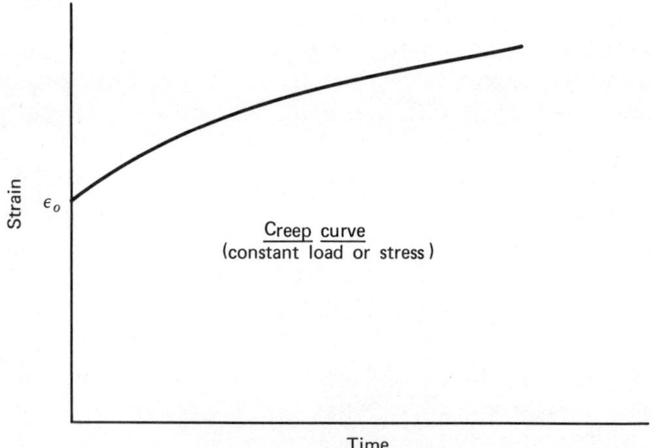

FIGURE 2.6 Creep curve.

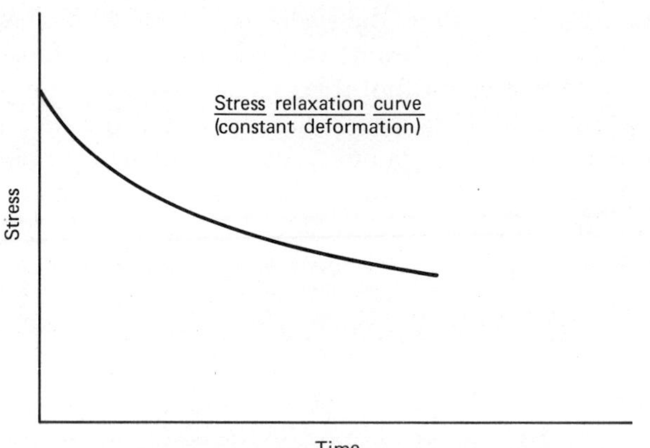

FIGURE 2.7 Stress relaxation curve.

composite member, the steel beam is virtually free of creep and stress relaxation. As the loads are applied, the member deflects a certain amount, and the stress relaxation in the concrete begins. As the loading is continued with time, the creep in the concrete affects the deflection so that the creep and stress relaxation affect each other. The net effect is a downward shift of the neutral axis. This shift raises the steel stresses slightly and decreases the effectiveness of the slab. There is, however, an increase of deflection with loads of long duration.

Even in bridge construction, which is generally unshored, there is some portion of the dead load which is not added until the slab has cured and the section has become composite. This load usually consists of a hand rail, curb, safety walk, light standards, and sometimes a separate wearing course on the roadway.

Neither AISC nor AASHTO specifies any computation of stresses under long-term duration of load. Knowles has shown that the value of creep stress at the bottom flange of a steel-concrete composite beam is about 12%. Most codes recognize this and consider it not enough to affect design stresses.

Long-term deflections, however, should be checked. AISC recommends a value of $2n$ for building design, and AASHTO recommends a value of $3n$ for computing these deflections.

It is interesting that the AISC value of $2n$ reduces the effectiveness of the slab to one-half of its previous value. Consequently, the partial slab tables in the AISC manual give a good estimate of the reduced section properties.

In concrete-concrete composite construction, the values of n for the beam and slab are close to each other so that the effect of creep in the composite member is virtually the same as in the monolithic T beam.

In timber-concrete construction, both the timber and concrete have creep properties. Furthermore, these properties, while not equal, are at least in the same order of magnitude so that effectively there is no shift of the neutral axis and no significant change in stress due to creep under long-term loading. Consequently the Timber Institute does not require a separate computation of long-term deflections.

2.6 Shrinkage and Temperature Change

The shrinkage of concrete as it cures sets up a differential strain between the slab and the beam, as the slab tends to bond to the shear connectors and to the top flange of the beam. As the concrete shrinks, tensile stresses are set up in the slab. This action puts the top of the steel beam into compression and the bottom flange of the steel beam into tension. Exact solutions of this problem have shown that the shrinkage stresses are not large enough to overstress seriously the steel beam. To check the shrinkage stresses for design purposes, a value of 0.0002 in. per inch of strain in the concrete is usually assumed. The shrinkage stresses may be large enough to crack the concrete, but positive moment loading which places the slab in compression tends to keep the cracks closed. The shrinkage does exert a force on the shear connectors, but the connectors in steel-concrete design are based on ultimate strength considerations which depend on the redistribution of load among the connectors, so that shrinkage is usually neglected in connector design.

Temperature differential between the beam and slab should be considered in some types of construction. In steel-concrete construction the coefficients of expansion of the steel and concrete are approximately equal, so that in building work there should be no temperature differential between the beam and slab. In bridge work, the top of the slab is exposed, whereas the beams are shaded. On a 90° sunny day, it is not unusual for slab surface temperature to reach 135°. The bottom of the slab, of course, is cooler than this, but it still may be well above the beam temperatures. When the top of the slab is warmer than the bottom, the slab action is somewhat analogous to the curl of pavement slabs, and it exerts some uplift force near midspan. However, at present there are no provisions required by any major code for extra connectors in steel-concrete construction due to temperature.

In timber-concrete composite construction the wood is assumed to be unaffected by temperature change. The concrete expands with heat, so the AITC specifications require a check for the number of additional shear connectors to account for the differential expansion. It is not unusual to find that the temperature change requires about the same number of connectors as the live load shear.

2.7 Loadings

In building work, most live loads are specified as uniformly distributed loads in pounds per square foot of floor area. With this type of loading in steel construction, shear connectors are uniformly spaced. However, there are cases where large concentrated loads are placed on the span. A typical case is shown in Figure 2.8. In this case, the AISC specifications require a separate computation for the number of shear connectors required between the concentrated load and the point-of-zero bending moment.

In bridge work, the live loads applied to the span are the AASHTO series H or HS loadings. The H and HS loadings consist of either truck load or lane load, whichever is critical. Truck loads are usually critical for short and intermediate spans and the lane load for longer spans.

The truck loading consists of one design truck placed on the span in such a position as to cause maximum bending moment and repositioned so as to cause maximum shear. Figure 2.9 shows an HS-20 truck.

The lane loading consists of both a uniform load and a movable concentration. The lane load is a simulation of a train of medium weight

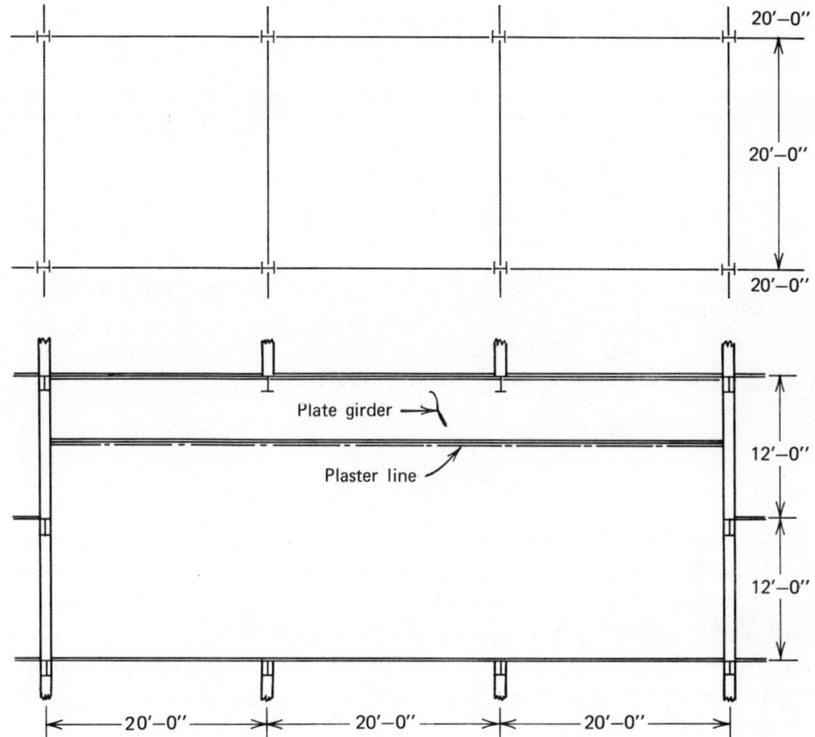

FIGURE 2.8 Concentrated loads on plate girder.

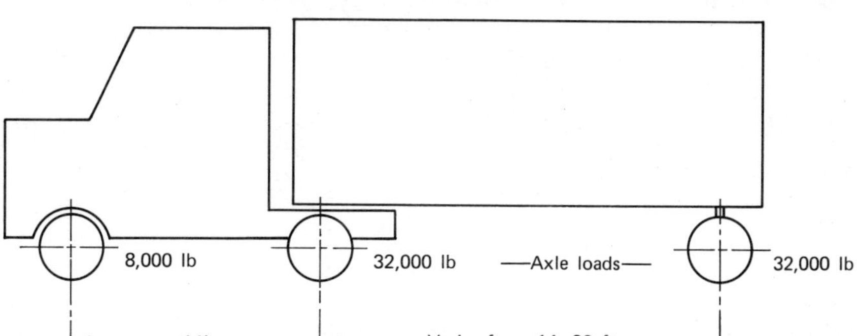

FIGURE 2.9 HS-20 truck.

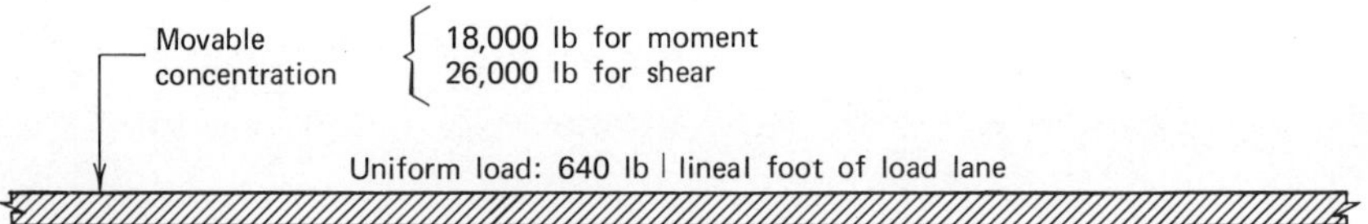

FIGURE 2.10 HS-20 lane loading.

vehicles with one heavy vehicle that can occupy any position on the span. Figure 2.10 shows the HS-20 loading.

According to the AASHTO specifications, shear connectors can be spaced either at regular or variable intervals.

2.8 Deflections

The composite beam is generally much stiffer than its noncomposite counterpart, so that for equal loads and spans, the deflections of the composite beam should be less.

Allowable deflections are generally based on conditions of the structure. In building work, one common limit on deflections is 1/360 of span length. This limitation prevents cracking of a plastered ceiling on the underside of the beam.

In bridge work, the deflection due to live load plus impact should not exceed 1/800 span. This limitation helps to provide a smooth riding bridge.

Some codes set up a slenderness limitation of beams, that is, the depth to span ratio. Under the ACI code, for example, if a beam is not supporting or attached to other building elements which might be damaged by large deflections, actual deflections may not have to be calculated. In an unshored composite member, if the overall thickness of the member exceeds $L/16$, deflections need not be computed. In a 24-ft span, then, if the overall depth of the composite T beam exceeds

$$h \geqslant \frac{L}{16}$$

$$\geqslant \frac{24 \times 12}{16} = 18 \text{ in.}$$

the beam is considered satisfactory for deflection. If the beam is shallower than this, then the deflections must be calculated and the allowable short-term deflection is the same $L/360$ used by other codes.

2.9 Construction

Most codes and books when referring to composite construction speak of shored and unshored construction. Shoring, in this sense, means that the beam member is supported from below so that the dead load of the concrete slab and the beam member are supported by the falsework. Actually, in steel-concrete construction in buildings, the beam is seldom shored. However, the slab forms are usually supported from below. This type of construction is considered unshored construction because the beam is free to deflect under dead load. It is usually cheaper and safer to support the slab from below than to design a relatively heavy system of slab form and walers to be hung from the beams. In bridge work, it is usually necessary to hang the forms from the beams. However, if slab supports are used in building work, the dead load of the concrete puts a deformation and stress into the falsework. As the concrete sets, it makes this deformation permanent and does not relax the stress. Consequently, the intermediate supports should have a provision for slowly relaxing this stress before removal.

In concrete work, it is more common to see a complete falsework system supporting both the beam and slab forms. Naturally, the same precautions should be taken to relax the stress in the falsework before removal.

References

2.1. Hover, K. C., "Elements of Falsework Design for Concrete Bridges," M.S. Thesis, Department of Civil Engineering, University of Cincinnati, 1974.

2.2. Cook, J. P., "Composite Construction Methods," *ASCE Construction Journal*, March 1976.

2.3. Hill, L. A., Jr., *Fundamentals of Structural Design*, Intext Publishers, New York, 1975.

3

Steel-Concrete Construction

The steel beam with a concrete slab is considered the standard for comparison in composite construction. Ever since composite construction was accepted by the major code-writing bodies, about 1944, most engineers and architects have been aware of the inherent advantages, such as the savings in steel, reduction in beam depth, increased stiffness, and increased overload capacity.

The deterioration of concrete bridge decks has been cited as one of the most critical problems facing structural engineers today. At least part of the solution to this problem lies in composite construction. In the noncomposite bridge, the top of the slab is in compression and the bottom is in tension. The tensile region of the concrete naturally cracks. These cracks are worked open and closed with the repeated application of live and impact loads so that they tend to grow. The composite bridge helps this problem by keeping the slab in compression.

One reason that composite construction has not achieved its full potential is that the selection of the steel section has been largely a time-consuming, trial-and-error process. Experience often led the designer to rough out a steel beam size based on no composite action. Then, knowing he would achieve about 20 to 30% in savings in steel weight, he could pick a trial section. Now, however, there are design aids available which shorten the process. Lothers (3.1) presents a method of using an increased allowable stress which makes a rapid selection of a trial section possible. The AISC manual presents tables of composite-section properties for steels of two different yield points, using the slab thicknesses which are prevalent in building design. In a later section of this chapter, one of the illustrative problems will be checked using these tables.

3.1 Types of Steel-Concrete Composite Beams

The most frequently used type of steel-concrete composite beam is the symmetrical, rolled steel, wide-flange section with a concrete slab. This type of structure is usually the fastest composite beam to design and build. Figure 3.1 shows this type of beam. The haunch shown in Figure 3.1*b* can be cast so that it encloses the top flange of the steel section. Although not accounted for in any calculations, this haunch concrete does help to provide some extra stability against the lateral buckling of the compression flange of the beam.

The reason for using composite construction is to utilize each of the component materials to its best advantage. The concrete slab is supposed to act as a big top coverplate for the rolled beam. This coverplate naturally forces the neutral axis of the composite member upward. In many cases, the neutral axis moves up so far that it falls within the slab. With some beam and slab combinations, an unsymmetrical beam makes a better design. This is easily accomplished by adding a coverplate to the bottom flange of the beam. This extra steel at the bottom of the section acts to move the neutral axis downward. Figure 3.2 shows this type of coverplated beam.

In the usual steel-concrete composite member, with a slab resting on the steel beam, the natural bond between the beam and slab is neglected in shear calculations. However, if the beam is fully encased in concrete and certain design cirteria are met, the encased beam can be considered as a composite member. Historically, one of the early uses of portland cement concrete was as a fireproofing material for steel beams, so that some of these early beams provided some measure of composite action. Figure 3.3 shows an encased beam.

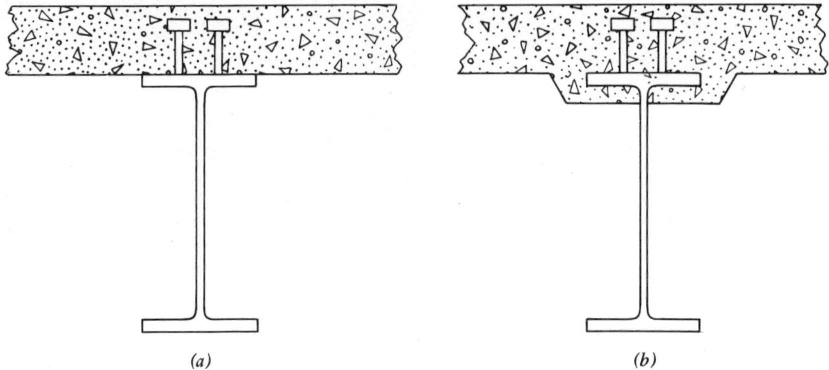

FIGURE 3.1

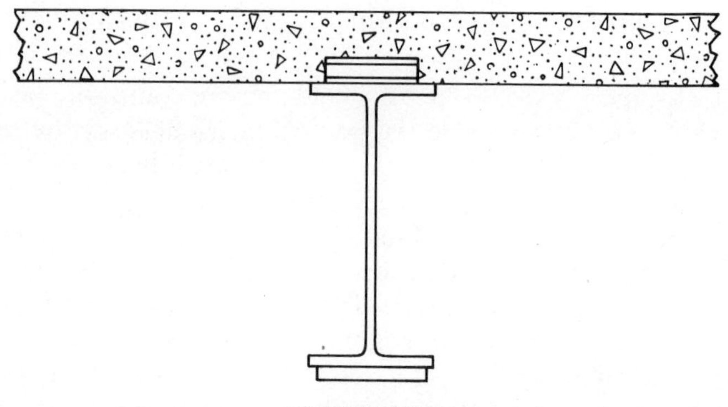

FIGURE 3.2

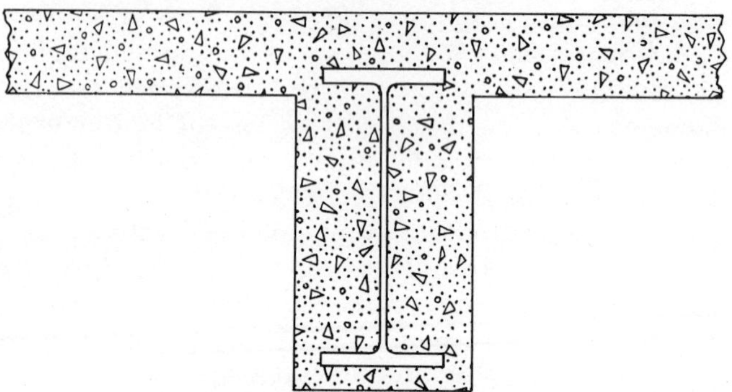

FIGURE 3.3 Encased beam.

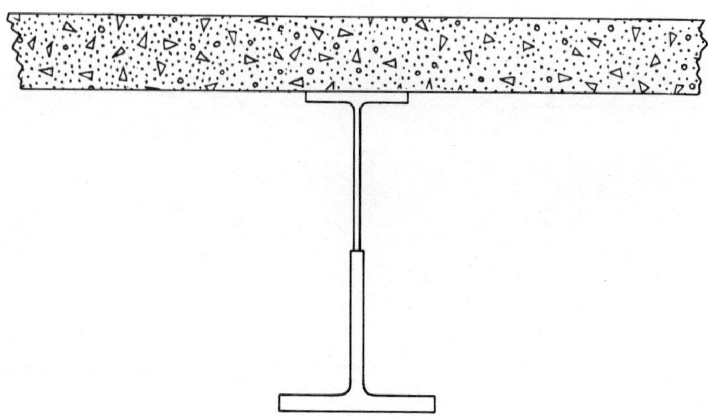

FIGURE 3.4 T-on-T beam.

In composite beam design, the neutral axis moves upward toward the slab. Consequently, the top flange of the steel beam is quite close to the neutral axis where it carries very little stress. The "T-on-T" composite member uses a large lower T. The upper T, which is in a low stress region, can then be much smaller. This type of structure saves steel poundage, but this savings must be balanced against the cost of welding the two T sections into a unit. Figure 3.4 shows a T-on-T beam. Figure 3.5 shows a composite member that carries the T concept to its obvious conclusion. A large T is used as the steel member. This member can be a rolled T or it can be made by welding two plates together.

Studs are welded to both sides of the top of the web. This member can be structurally very efficient, but it does pose some construction problems. It is quite normal in steel and concrete construction to use the top flange of the steel member to support the slab forms. This member has no top flange, so the formwork must be carefully designed and detailed (Fig. 3.6). Also, in the design of composite members, the steel section alone must support the dead load of the formwork and the wet concrete. Composite action does not begin until the concrete has hardened. In steel beam design, lateral buckling of the compression area of the beam may be critical. Along with the conditions of lateral support, the dimensions of the compression flange of the beam can determine the allowable bending stress. This beam has no compression flange. Consequently, sufficient lateral support must be provided to prevent buckling at the top of the web until the concrete has cured. If this beam can be fully shored from below until the concrete has hardened, then both dead and live loads are carried by the composite beam.

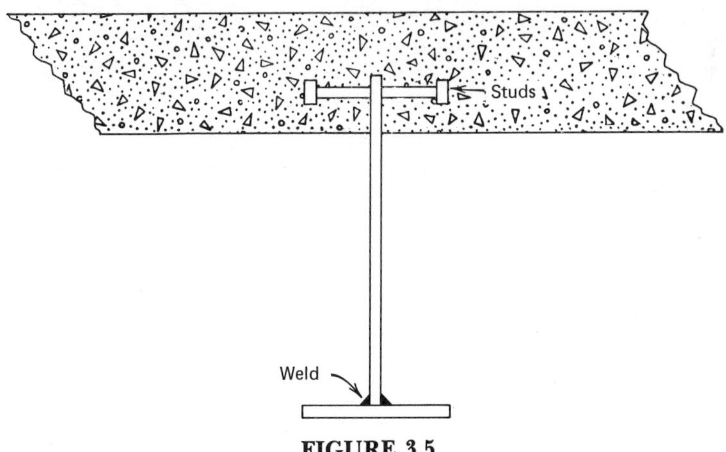

FIGURE 3.5

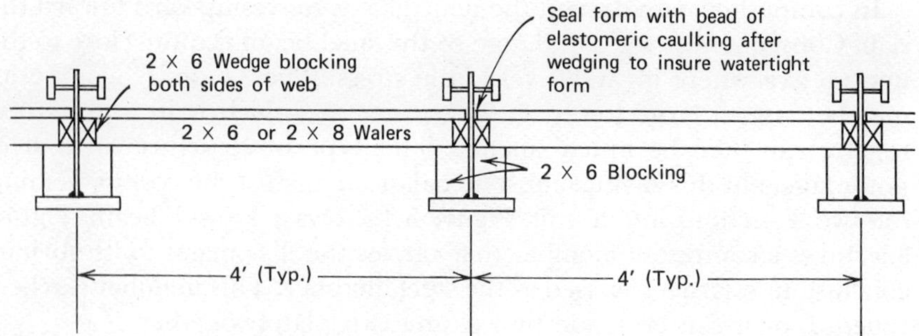

FIGURE 3.6 Formwork for flangeless beam.

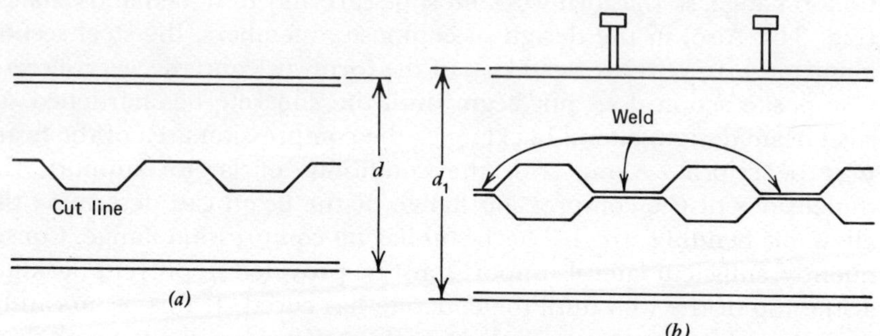

FIGURE 3.7 Castellated beam.

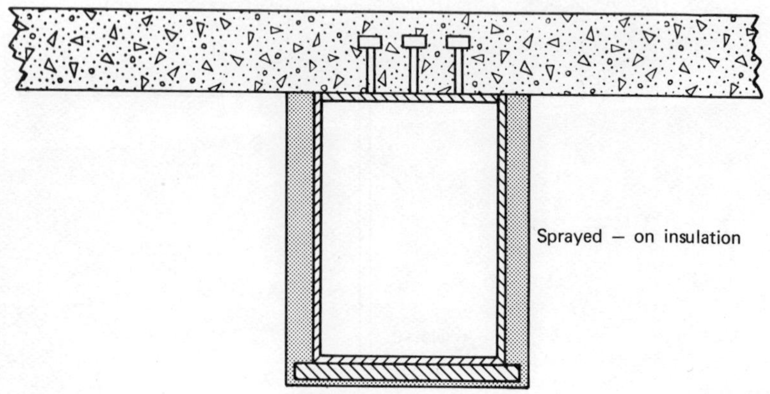

FIGURE 3.8 Duct beam.

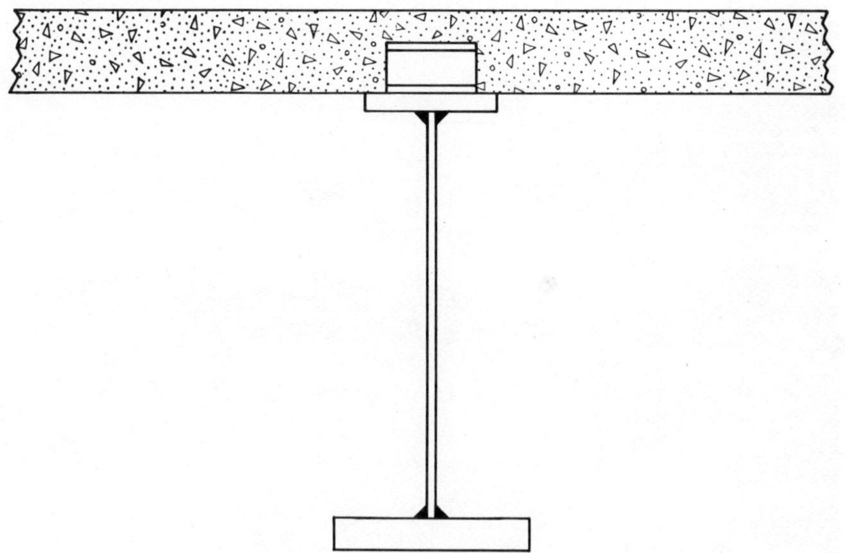

FIGURE 3.9 Composite plate girder.

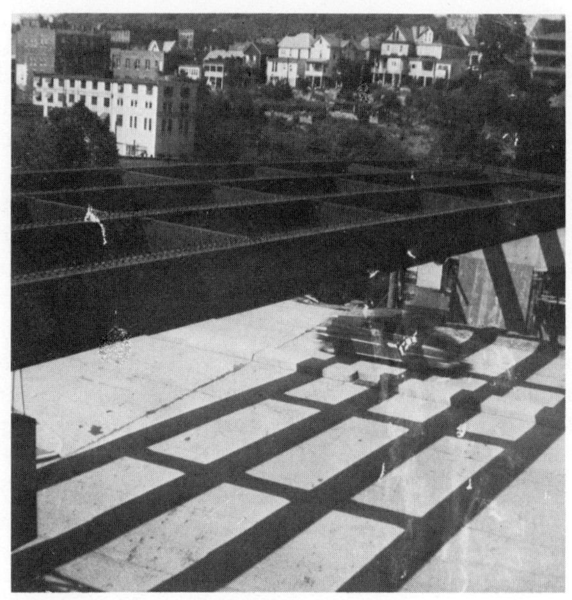

FIGURE 3.10 Second Street bridge, Clarksburg, West Virginia. (Courtesy of Frank D. McEnteer.)

FIGURE 3.11 Second Street bridge, Clarksburg, West Virginia. (Courtesy of Frank D. McEnteer.)

The castellated beam shown in Figure 3.7 is made from a single rolled section. The beam is shop cut and then welded together again to form a deeper section. In addition to its greater depth, this beam provides void space for piping and electrical conduit. The box beam provides good torsional rigidity as well as load-carrying capacity. Composite box beams are being used more and more frequently in long-span bridges. One enterprising design firm* has used the box beam in buildings and utilized the space as the main header for the heating and cooling system of the building. Figure 3.8 shows a section of this duct beam. The plate girder can be fabricated unsymmetrically with a larger tension flange so that it forms an excellent composite beam (Fig. 3.9). The bridge shown in Figures 3.10 and 3.11 is a simple span, composite plate girder. This is an interchange bridge crossing a main artery. Construction traffic had to be maintained on the roadway below, so that shoring was impossible. This bridge was built in 1957. Figure 3.10 shows the spirals that were still widely used at that time.

*Reid & Tarics, Inc.

3.2 Construction Methods

The usual discussion of construction methods in composite design begins and ends with shoring. However, there are many other factors besides shoring which should be considered.

3.2.1 Shored or Unshored

After the steel beams are erected, the concrete slab is cast. The sequence of slab casting is discussed in a later section. The slab forms, wet concrete, and the temporary construction loads must either be shored from below or hung from the steel beam. If no shoring is used, the steel beams support all the dead loads, including the weight of the beams themselves. When shoring is used, the falsework supports the load of the concrete and other construction loads. After the slab has cured and the shores are removed, both dead and live loads are carried by the composite section.

At working loads there are initially different stress distributions in shored and unshored beams. These are shown in Figure 3:12. Assuming equal loads, the slab stresses in the shored beam are higher and the steel stresses are lower than in the unshored beam. However, in the shored beam the slab carries the dead load stresses which are long-term effects, so that dead load creep acts to change this pattern. The effect of creep and stress relaxation is to shift stresses out of the slab into the beam, so that in the long term, the stress patterns for the shored and unshored cases are very nearly equal. Tests have shown that the ultimate carrying capacity of composite beams are the same whether shored or not. AISC

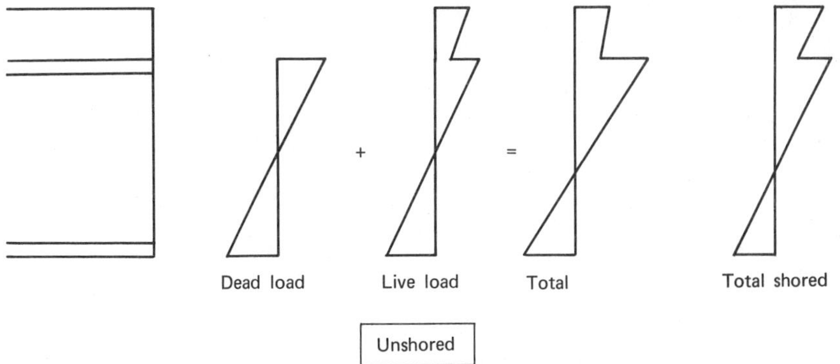

FIGURE 3.12 Stress distributions.

Table 3.1 Partial List of Major Falsework Failures

Year	Project	Nature of Failure	Probable Cause	Killed	Injured	Ref.
1931	U. of Cal. ME Bldg., Berkeley, Calif.	Roof slab collapsed	Poor bracing	2	10	M
1937	Chapel Tower, Duluth, Minn.	Floor slab collapsed	Inadequate vertical support	4	2 +	M
1937	Allentown Reservoir, Pa.	Roof slab collapsed	Unknown	0	0	M
1942	Water Reservoir, Washington, D.C.	Roof slab collapsed	Unknown	0	13	M
1949	Water Reservoir, Santa Monica, Calif.	Roof slab collapsed	Shores failed	1	2 +	P
1949	Park-Bernet Galleries, New York, N.Y.	Floor slab collapsed	Shores failed	0	15	P
1953	Esso Oil Co. Reservoir, Everett, Mass.	Roof slab collapsed	Shores failed	1	18	P
1953	Concrete Bldg., Scarsdale, N.Y.	Floor slab collapsed	Inadequate reshores	3	10	P
1955	Municipal Coliseum, New York, N.Y.	Floor slab collapsed	Dynamic forces	1	50	M
1955	State Fair Stadium, Louisville, Ky.	Floor slab collapsed	Shores failed	0	7	P
1959	Georgia Power Co. Office, Atlanta, Ga.	Floor slab collapsed	Shores failed	1	15	P
1959	Johnson & Johnson, Ltd., Montreal, Canada	Roof slab collapsed	Unknown	4	20	P
1960	Smithsonian Inst., Washington, D.C.	Floor slab collapsed	Inadequate bracing	0	12	P
1960	Municipal Parking Garage, Newark, N.J.	Roof slab collapsed	Improper reshores	0	0	M
1960	Apartment Bldg., Arlington, Va.	Floor slab collapsed	Commercial shore failed	1	4	P

1961	United California Bank Bldg.	Roof slab collapsed	Unknown	0	0	P
1961	Toronto Subway, Canada	Roof slab collapsed	Shores failed	2	0	P
1971	Concrete Bridge, Sacramento, Calif.	Falsework collapsed	Unknown	2 +	2 +	A
1971	South Mojave Overhead, Calif.	Falsework Tower blown over	Wind load	0	0	A
1971	Railroad Underpass, Ventura, Calif.	Deck form collapsed	Unknown	0	2 +	A
1971	Post-Tensioned Bridge, Baldwin Park, Calif.	Falsework collapsed	Unknown	0	0	A
1971	Concrete Bridge, Oceanside, Calif.	Falsework Towers collapsed	Unknown	2 +	2 +	A
1972	Concrete Bridge, Pasadena, Calif.	Deck collapsed	Instability	6	6	E
1972	Overpass, Dallas, Texas	Deckform fell	Defective hangers	0	2	E
1972	Concrete Bridge, Koblenz, W. Germany	Conc. girders collapsed	Unknown	6	15	E
1972	Concrete Bridge, London, England	Deck collapsed	Falsework design error	3	10	E
1973	Hi-Rise Apt., Bailey's Crossroads, Va.	Multiple floor collapse	Early shore removal	14	30	E

Reference Code: A = ASCE; E = ENR; P = Peurifoy; M = McKaig.

states that "the composite section shall be proportioned to support all of the loads . . . even when the steel beam is not shored during construction."

So the practical question remains, "Should the beam be shored or not?" In bridge work, the answer is usually "no," simply because there is no adequate base on which to support the shores. Also, shoring is tricky. It involves the use of many slender compression members. If the base washes out from under one of these, the whole system could collapse. In building work, shores can be adequately supported by the slab below. Shored construction does provide a smaller steel section. However, the additional cost of shoring usually nullifies this advantage.

Actually, a beam need not be continuously shored throughout its length. In building work, most design loads are uniformly distributed. Since the bending moment varies as the square of the span length, even one sturdy central shore which cuts the span in half would be a considerable advantage.

However, shoring and shoring removal is a tricky business and the following table from Hover (2.1) would seem to present a rather compelling case against shoring.

If the beam is to be shored, the engineer must set the falsework carefully to maintain the proper slab thickness. The slab must act as a top coverplate for the beam, so its thickness is important. If the shores are forced in too tightly at midspan, giving the beam an upward deflection, the slab, when screeded level, may be thinner than the design slab.

Considering all the factors involved, the average designer is probably better off sticking to unshored construction for most cases.

If the beam is not to be shored, the engineer should check the deflection of the steel beam alone, under the dead load of formwork, wet concrete, rebars, and the men and equipment placing the concrete. It is possible under those temporary construction loads to get an excessive midspan deflection, which means that more concrete must be added to bring the slab up to grade. This is added dead load and a slab which is thicker than the design at the section of maximum positive bending moment.

3.2.2 Casting Sequence of the Concrete

In continuous beam bridge construction, the placing sequence of the deck concrete takes on a particular importance. Figure 3.13 shows a plan of a typical three-span composite bridge. The concrete placing sequence is indicated by (1) and (2). This placing sequence assumes that ready-mix

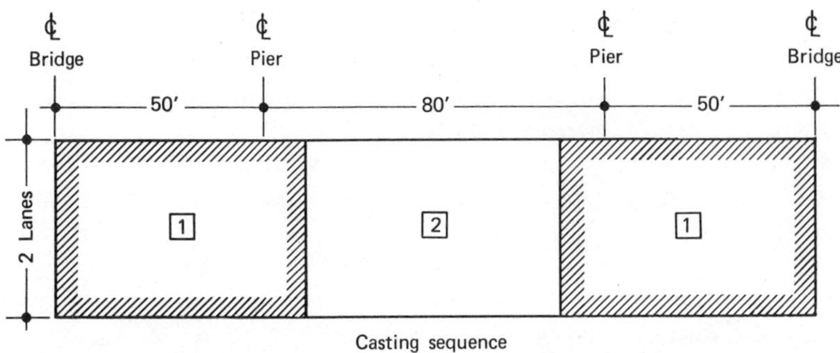

FIGURE 3.13 Slab casting sequence—three-span bridge (C_l = centerline).

trucks can be brought to both ends of the bridge, and that the concrete will be placed by buggies. Figure 3.14 shows the dead load on the structure and the deflected shape of the structure after the completion of Pour No. 1. This concrete takes its initial set and bonds to the shear connectors with the structure in this deflected position.

Naturally, the construction crews want to proceed with the concrete placement as soon as practicable. When Pour No. 2 is placed, the first concrete placement is only a few days old. When the second batch of concrete is placed in the center span, the bridge returns to its intended level position. However, it is possible that this construction sequence can break the bond between the shear connectors and the green concrete in Pour No. 1. This would result in some slip and incomplete composite action between beam and slab. To solve this problem, one state highway department planned continuous placement of the concrete from one end of the structure to the other. Since the bridge involved was rather long, the engineers used a retarding agent in the concrete, and varied

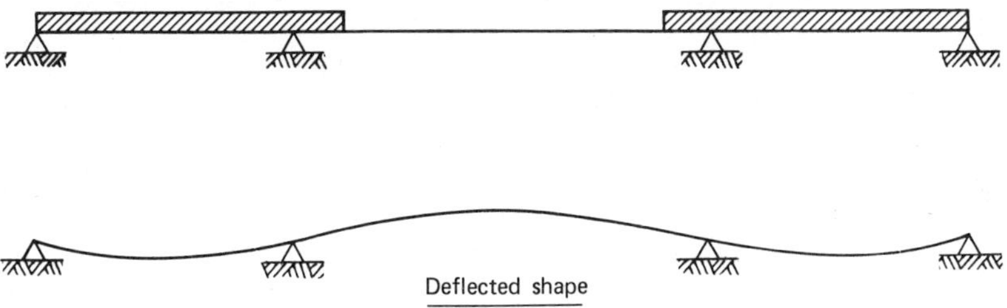

FIGURE 3.14 Deflected shape after Pour No. 1.

the percentage of retarder throughout the span. In this way, all the deck concrete reached its design strength at approximately the same time.

The designer can use the placing sequence of the concrete to work in his favor. Knowles (1.2) outlined a method of "phased casting," by which about half of the slab dead load bending moment is transferred from the steel beam to the composite section. The sequence of operations is as follows:

1. Erect the steel beams.
2. Cast the middle portion of the slab.
3. After the central portion of the slab has hardened and become composite, the remaining sections of the slab are cast.

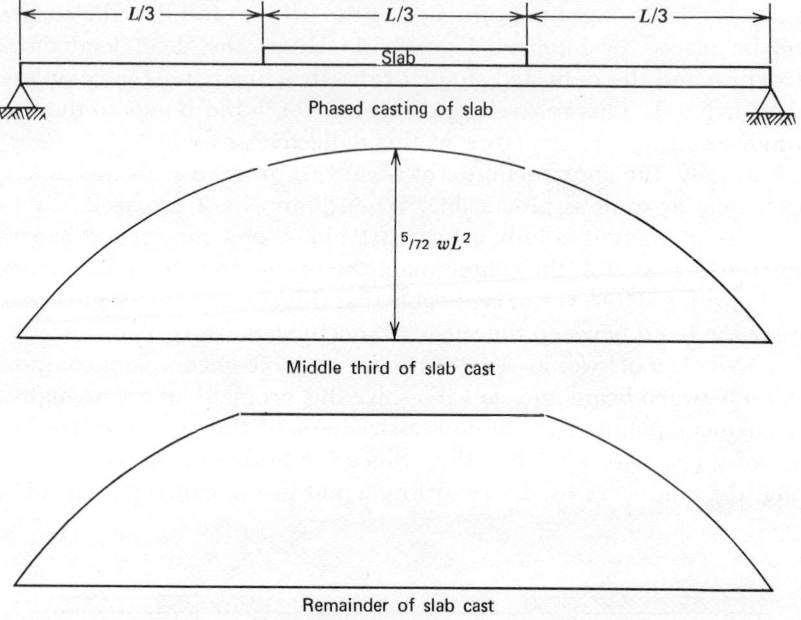

FIGURE 3.15 Phased casting.

If the middle third of the slab is used, the reduction amounts to 44% of the slab dead-load moment. If the middle quarter of the slab is used, the reduction is above 50%. This, however, is only a reduction in the slab dead-load bending moment. If the live-load to dead-load ratio is high, or if concrete placement is difficult, this method may not be worth the extra cost, but under the right job conditions it can pay valuable dividends.

3.2.3 Use of Sheet Metal Forms

The steel deck composite beams are usually associated with building work, where the steel decking can provide raceways for electrical and telephone services. However, "stay-in-place" metal forms are gaining wider acceptance in bridge construction also. Figure 3.16 shows the studs and sheet metal formwork for a bridge in New York. Figure 3.17 shows another bridge with steel decking. Studs and reinforcing bars are in place, ready for the concrete deck.

For short span bridges, the county engineer in Montgomery County, Ohio, has revived the old "jack arch" construction that used to be popular for floor systems in buildings. The system, shown in Figure 3.18, uses closely spaced beams and sheet metal forms which are readily available and easy to erect. No shear connectors are used. Strictly speaking, this is not an encased beam by AISC standards because the entire beam is not encased in concrete. However, most of the beam is encased, and a quick check of the section properties shows that the moment of inertia is doubled and the section modulus is increased by about 35%. This type of construction should provide very effective composite action.

FIGURE 3.16 Sheet metal forms. (Courtesy of Structures Subdivision—NYS DOT.)

FIGURE 3.17 Ready for concrete. (Courtesy of Structures Subdivision—NYS DOT.)

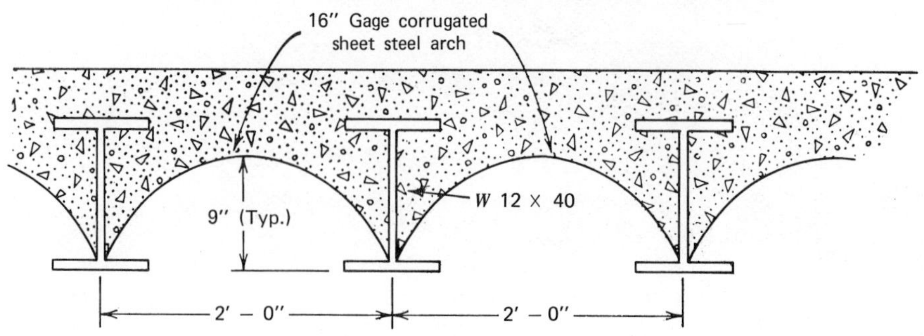

FIGURE 3.18 Partially encased beam.

3.3 Proportioning of Composite Members

In unshored construction, the bare steel beams must be proportioned to carry the dead load of formwork, wet concrete, and the temporary construction loads of the men and equipment placing the concrete.

In shored construction, the composite section is assumed to carry all the loads.

When the beam is in service, the composite beam must be designed to carry the full dead and live loads.

However, just as in any other steel design, other considerations of lateral support, deflections, and the use of coverplates affect the final proportions of the composite member.

3.3.1 Effective Flange width

The steel beam with a concrete slab is effectively a T beam. The portion of the slab directly above the steel section, which is bonded to the steel section, naturally contributes most fully to the composite action.

As you move away from the steel beam, the slab becomes less effective in carrying the longitudinal stresses. Consequently, the width of the slab which is considered effective is limited by specification. Referring to Figure 3.19, the AISC specifications limit the value of "b" to the least of the three following values:

1. b not greater than one-fourth of span length.

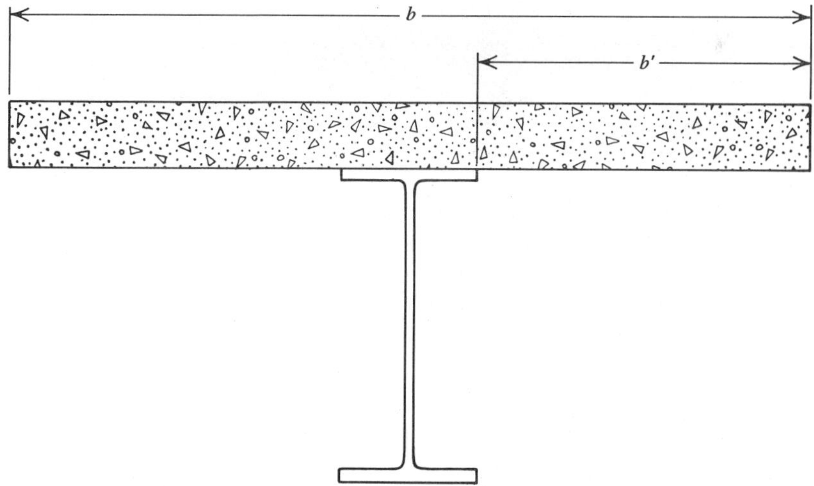

FIGURE 3.19 Effective flange width.

2. b' not greater than one-half the clear distance to the adjacent beam.
3. b' not greater than eight times the slab thickness.

Practically speaking then, the effective flange width is the least of three values.

1. $b \leqslant \frac{1}{4}$ span.
2. $b \leqslant c/c$ beams.
3. $b \leqslant 16t + b_f$.

If the flange exists on only one side of the beam, the requirements are somewhat different (see Fig. 3.20).

1. $b' \leqslant$ one-twelfth of span.
2. $b' \leqslant$ one-half the clear distance.
3. $b' \leqslant$ six times the slab thickness.

Under the AASHTO specifications for bridges, the first two requirements remain the same, but the requirement relating to slab thickness is much more conservative. This limits the effective flange width to 12 times the slab thickness.

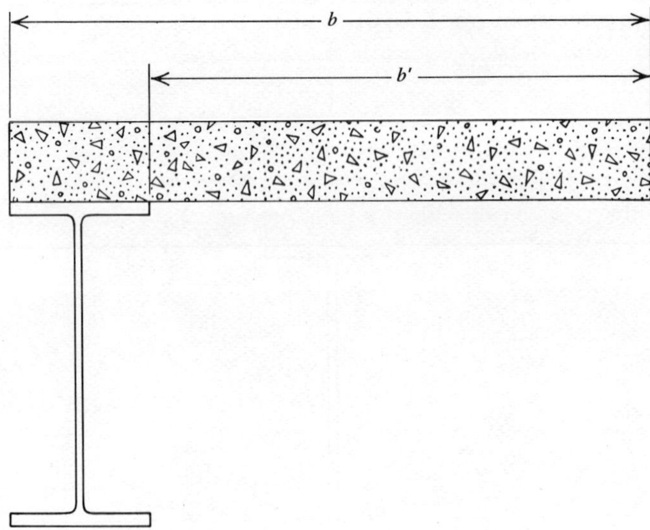

FIGURE 3.20 Flange on one side only.

3.3.2 Negative Bending

The design of composite beams for negative bending is permitted by both AISC and AASHTO specifications. However, the question of negative bending is only a portion of the larger question of simple versus continuous spans. With continuous construction, bending moments are lower so that smaller steel sections are required. In a typical three-span continuous bridge, the highest bending moments are over the two interior supports. Consequently, the greatest economy is usually realized by selecting a steel section to carry the positive moments and adding coverplates in the negative regions over the supports. However, these maximum negative moments in the continuous beam are lower than the positive moments in the simple span beam with equal span lengths.

With continuous construction, there are other costs which must be considered in addition to the cost of the steel. In a three-span bridge, continuous construction is more efficient when the center span is longer than the side spans. The usual method of construction is to erect the side span steel, which cantilevers out into the center span to approximately the dead load point of contraflexure. The center span is then field bolted into position. This field splice adds to the cost of continuous construction. However, the three-span continuous beam requires only four bearing shoes compared to the six required for the bridge of three simple spans. The continuous bridge also has only two deck expansion joints that must be sealed, whereas the simple span structure has four.

The composite bridge benefits from the same advantages as the noncomposite bridge in regard to the number of shoes and expansion joints.

In composite construction, the concrete slab in the negative moment regions is considered cracked and makes no contribution to composite action. However, even though the slab is cracked, it is still capable of transferring load to the reinforcing bars. Thus the composite section for negative bending consists of the steel beam and the area of the longitudinal bars in the slab. Both AISC and AASHTO provide a formula for determining the number of connectors in the negative regions. These formulas are based on the area and strength of the reinforcing bars.

The cost comparison shown below is based on a three-span bridge with equal spans of 60 ft. This is not the most economical span ratio, but it is used here to simplify the computations. These figures include the cost of studs for the composite beam, coverplate welding for the plated beam, bolted splices in the continuous beams, and the cost of extra bearing shoes and expansion joints for the simple span beams. The cost shown in each case is the cost of one beam. Slab, reinforcing, formwork, and other factors which are largely constant for all the sections are not included.

Simple beams, noncomposite	W 30 × 108	\$6252
Simple beams, composite	W 27 × 84 with studs	\$5233
Continuous beams, noncomposite	W 27 × 94 with coverplates	\$5608
Continuous beams, composite	W 24 × 76 with studs	\$4500

Deflections of these four beams vary with the method of construction and the stiffness of the cross section. Composite and continuous constructions both reduce the deflections. The deflections shown are live load deflections and are based on a uniformly distributed load that is roughly equivalent to AASHTO loading for this span length.

Simple beam, noncomposite	Deflection = 1.45 in.
Simple beam, composite	Deflection = 0.82 in.
Continuous beam, noncomposite	Deflection = 0.63 in.
Continuous beam, composite	Deflection = 0.50 in.

For building work, the same type of cost comparison can be made. Construction sequence would naturally be different, with continuity being achieved at the beam ends by moment-resistant connections. Of course, the extra expansion joints, bearing shoes, and field bolted splices would be eliminated from the comparison.

Simply because the use of shear connectors is permitted in negative bending regions, it should not be assumed that this is necessarily the best solution. Some designers feel that it is not economically feasible to attempt full composite action in negative bending. There is justification for this opinion. Consider a simple example:

Using a 4 × 60-in. concrete slab, $4 \times 60/9 = 26.22$ in.^{2} of equivalent steel is lost in the negative region. Using No. 10 bars at 4-in. spacing, only 19 in.^{2} of steel is provided.

Some engineers prefer to reduce or prevent tensile cracking of the slab by insuring noncomposite action in the negative moment regions and providing shear connectors in the positive region only. Insuring noncomposite action is easy to accomplish at the job site. A sheet of polyethylene film or building paper can be taped to the top flange of the beam in the negative region before casting the slab.

Another method of handling the negative moment regions is to prestress the slab in order to keep the slab in compression. There are various ways in which this prestressing can be done. The negative moment region of the slab can be cast and prestressed directly with cables (Fig. 3.21).

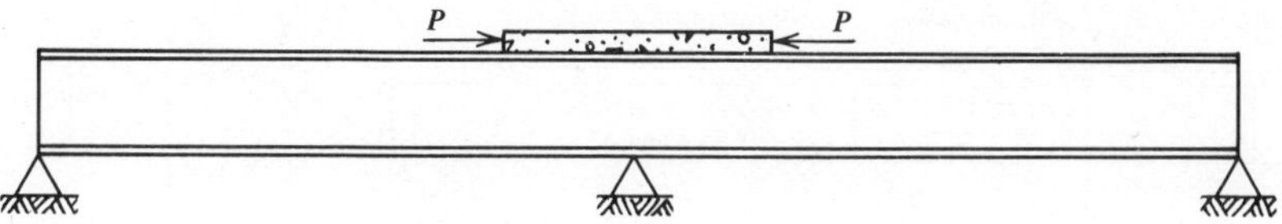

FIGURE 3.21 Slab prestressing.

The prestressing can also be done by planning the placing sequence of the concrete slab and jacking the supports. This latter method (1.2) was used on the Aboshi Bridge in Japan. This was a continuous beam with three spans of 105 ft each. The ends of the bridge at the two bridge seats were lowered about 2 in. The intermediate supports over the piers were jacked upward about 16 in. This placed the bottom flanges of the girders in compression and the top flanges in tension. After the slab was cast and cured, the structure was jacked back to its level position which placed the slab in compression. Although the prestressing method is used in building work, it is best suited to bridge work where the long spans and impact loads make these sophisticated techniques economically feasible.

3.3.3 Lateral Support

In ordinary steel beam design, the compression flange must be given adequate support in order to prevent lateral buckling of the compression flange. If lateral support is not provided, a reduced value of allowable stress must be used.

In composite construction, the slab that is attached to the beam furnishes this support. However, this lateral support is not effective until the slab has achieved 75% of its required strength.

In unshored construction in which the forms are hung from the steel beams, ordinary formwork does not provide adequate lateral support. The formwork should be designed to provide positive support for the compression flange if it is required. Figure 3.22 shows a section of the deck formwork for a bridge across the Ohio River, which is fairly typical for noncomposite construction. The walers, in this case double 2×12 members, are adequate to support the deck concrete, but are not intended to provide any lateral support for the beam.

In many composite designs the live-load to dead-load ratio is such that the dead load stresses in the beam are quite low. However, the reduced allowable stress for the unbraced compression flange should always be checked. This check becomes especially important if the governing code requires the application of a temporary construction live load. One of

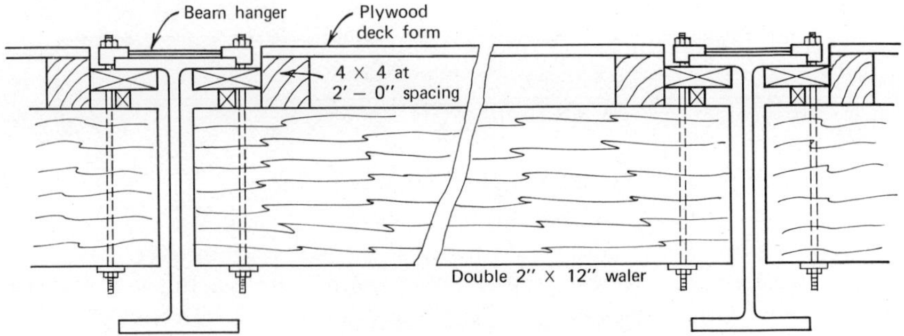

FIGURE 3.22 Slab formwork.

the illustrative problems in a later section of this chapter is checked for this reduced value of allowable stress.

For construction using encased beams, the side forms are at least 2 in. from the outside edge of the beam flange and do not provide any lateral support. In this type of construction particular care must be given to providing lateral support for the beam flange. One way this can be done is shown in Figure 3.23. Heavy reinforcing bars (Nos. 8, 9, or 10) are fillet

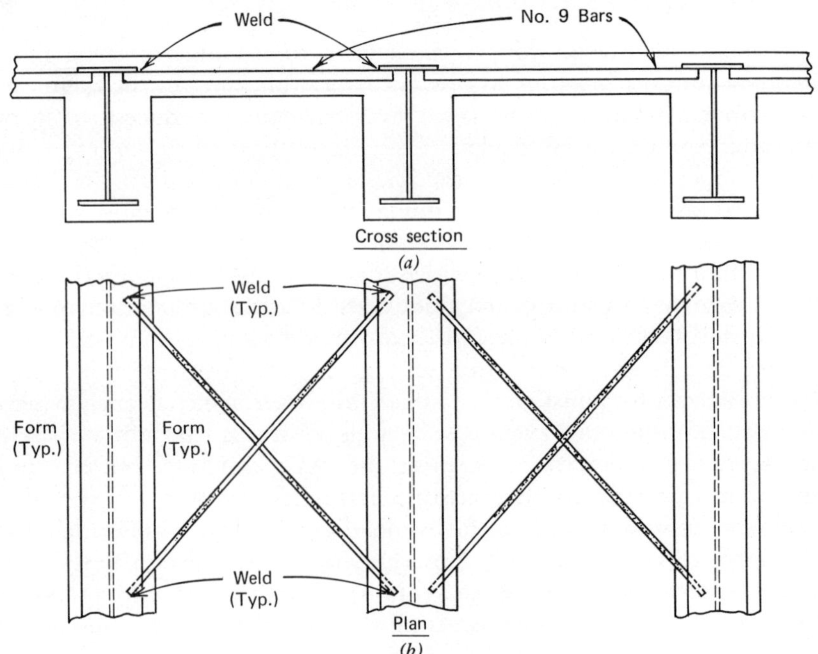

FIGURE 3.23 Lateral bracing.

welded to the lower side of the top flange of the steel beam in an X pattern. Note that the top flange of the beam must be far enough above the bottom of the slab form so that these braces can be encased in concrete and remain in the structure.

An alternate method is to prefabricate the transverse steel into a mat, place it on top of the beam flange, and tack weld each bar to the flange. This provides bracing at the bar spacing interval.

3.3.4 Deflections

Composite beams should be checked for deflection during erection and also for short-term and long-term deflection of the composite member under full load.

Deflection of the steel beam during erection can be important in maintaining the proper thickness of the deck slab.

Deflection of the composite member is checked for short-term loading using the modular ratio, n, in computing the moment of inertia. Long-term deflection recognizes the effect of creep in the concrete which transfers stresses out of the concrete into the steel beam. This computation uses a value of kn that gives a reduced value of moment of inertia and consequently larger deflections.

3.3.5 Coverplates

Coverplating of beams is often used in composite construction. In many cases of a symmetrical beam with a slab, the neutral axis of the composite beam falls within the slab. This is not the most efficient use of material because only part of the slab is acting in compression. The addition of a lower flange coverplate tends to move the neutral axis downward, usually to a point below the slab.

Cost figures must be checked carefully at this point. In a building with uniformly distributed load, plates can be cut off when they are no longer needed. As a rule of thumb, the coverplate length is usually about 60 to 70% of the beam span length. On this basis, adding a coverplate may save 20% of steel weight when compared to the alternative of using a larger symmetrical section. This savings in weight must be balanced against the cost of fabricating the plated beam. It costs virtually the same amount to weld on a thick plate as a thin one. So a general rule might be to use good size coverplates or none at all.

AISC recommends the following: if the coverplated section saves less than 7 ppf, do not use plates. If the coverplated section saves more than 12 ppf, use coverplates. Between these limits plates may or may not be economical depending on local fabricating costs.

The theoretical point of cutoff for coverplates can be determined either mathematically or graphically. The plate must extend beyond the theoretical point, and the extended portion must be attached with enough rivets, bolts, or weld length to develop the full strength of the plate.

The mathematical method of cutoff uses a simple parabolic relationship. Figure 3.24 shows the moment diagram for a uniformly loaded plated beam. The cutoff point, x, is determined from the following ratio: $x^2/(L/2)^2 = b/y$. Graphically, the moment diagram can be accurately laid out to scale. The resisting moment of the section with the plate and the section without the plate can be superimposed on the same drawing. The cutoff point is determined by the point at which the M_R of the unplated section intersects the moment diagram. Figure 3.25 illustrates the method.

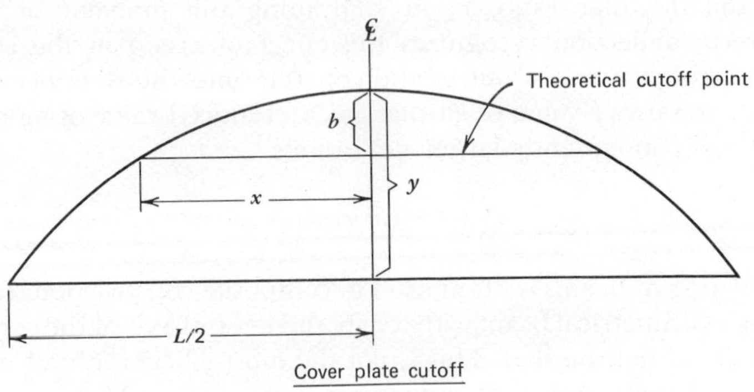

FIGURE 3.24 Coverplate cutoff.

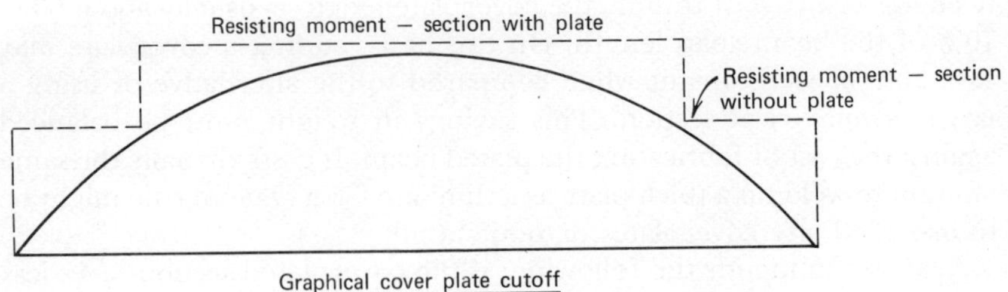

FIGURE 3.25 Coverplate cutoff graphical.

3.4 The Encased Beam

Steel beams in buildings can be encased in concrete for fire protection. Under some circumstances these encased beams can be designed as composite beams. No shear connectors, in the usual sense, are considered in this type of composite beam. The horizontal shear is transmitted from the steel beam to the concrete by friction and bond. In order to qualify as a composite beam, the concrete encasement must have mesh reinforcement throughout the whole depth and across the soffit of the steel beam, to prevent spalling of the concrete. The concrete encasement must also meet the minimum cover requirements shown in Figure 3.26. The slab and the encasement must be cast integrally.

There are advantages to the encased beam which are usually overlooked. Only the compression area of the concrete is considered in the design. However, all of the concrete contributes a valuable reserve of shear strength to the beam. Also, the shrewd designer will utilize the additional stiffness of the encased beam when analyzing the building for combinations of vertical and lateral loads.

Encased beams are not usually shored because there is no practical method of running the props through the bottom of the form to support the steel beam. The forms are generally hung from the beams, and the beams are designed as unshored members. Two-inch-high reinforcing bar chairs wired to the bottom of the beam form can be used to insure

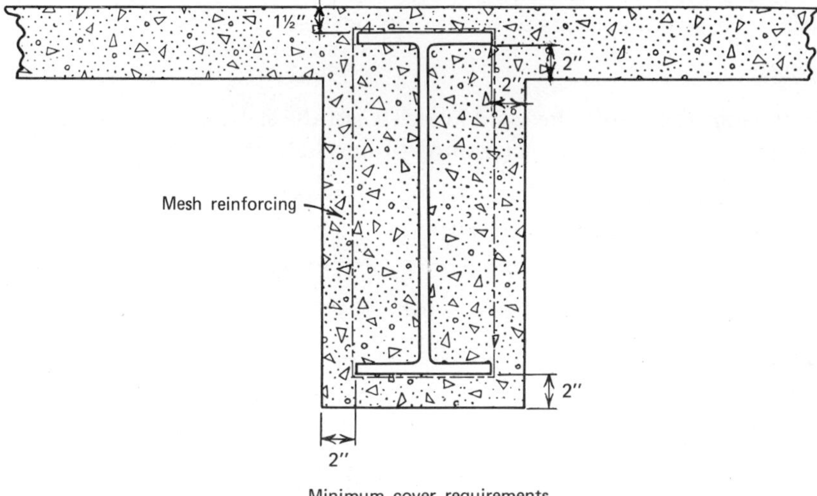

FIGURE 3.26 Encased beam minimum cover.

that the AISC minimum cover requirement is met. Two-inch cubes of concrete have been used for this purpose, but they have not been as satisfactory as other methods.

3.4.1 Design of the Encased Beam

The American Institute of Steel Construction allows an alternate design method for the proportioning of fully encased members. Instead of using the transformed section properties of the composite member, "the steel beam alone may be proportioned to resist unassisted the positive moment produced by all loads, live and dead, using a bending stress equal to $0.76F_y$, in which case temporary shoring is not required." This alternate method is illustrated in Example 3.1:

Example 3.1 The Encased Beam

BEAM SPACING	8 ft
SLAB "*t*"	4 in.
f'_c	3000 psi
SPAN	24 ft

$n = 9$
A-36 steel
$f_c = 1350$ psi

First use the AISC alternate design method with $0.76\ F_y$, then check using the actual composite section properties.

Loads

In the encased beam, the dead load (DL) of the slab and the encasement usually amount to about one-third to one-half of the live load (LL).

LIVE LOAD	150 psf
CEILING AND APPLIED FLOOR FINISH	25 psf
	175 psf

Assume dead load of concrete $= 0.4 \times 175 = 70$ psf.

DEAD LOAD	$70 \times 8 = 560$ ppf of beam
LIVE LOAD	$175 \times 8 = 1400$ ppf of beam
TOTAL LOAD	$= 1960$ ppf

$$M_T = \frac{1.96 \times \overline{24}^2}{8} = 141.12 \text{ kip-ft}$$

ALLOWABLE STRESS $\quad 0.76 \times 36 = 27.4$ psi.

Section Modulus Required

$$S = \frac{M}{f} = \frac{141.12 \times 12}{27.4} = 61.80 \text{ in.}^3 \text{ required}$$

$$\text{(try W } 16 \times 40) \qquad S = 64.6 \text{ in.}^3 \text{ supplied}$$

Check dead load of encasement. AISC specifications require 2 in. of clear concrete cover in the stem around the W section. The top of the steel beam must be at least $1\frac{1}{2}$ in. below the top of the slab. These requirements establish the stem dimensions shown in Figure 3.27.

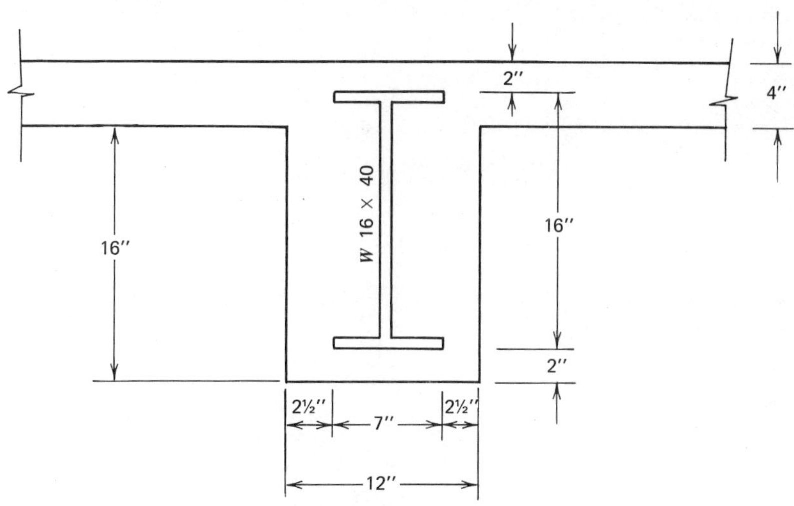

FIGURE 3.27

$$\text{Concrete area} = \text{Gross area} - \text{steel beam area}$$

SLAB	$4 \times 96 = 384 \text{ in.}^2$
STEM	$12 \times 16 = 192 \text{ in.}^2$
GROSS AREA	$= 576 \text{ in.}^2$
	Area of W $16 \times 40 = 11.8 \text{ in.}^2$
NET AREA	$576 - 11.08 = 564.2 \text{ in.}^2$
CONCRETE DEAD LOAD	$\frac{564.2}{144} \times 150 \text{ pcf} = 588 \text{ ppf.}$
STEEL BEAM	$= 40 \text{ ppf}$
TOTAL DEAD LOAD	$= 628 \text{ ppf}$

$$M_D = \frac{(628)(24)^2}{(1000)(8)} = 45.2 \text{ kip-ft}$$

$$M_L = \frac{(1400)(24)^2}{(1000)(8)} = \underline{100.8 \text{ kip-ft}}$$

$$M_T = M_D + M_L = 146 \text{ kip-ft}$$

ACTUAL STEEL STRESS $\quad \dfrac{146 \times 12}{64.6} = 27.12 \text{ ksi} < 27.4 \text{ ksi.}$

W 16 × 40 looks satisfactory and efficient.

Now check this result against the actual composite section properties.

Effective Flange Width

$$1/4 \text{ span length} = \frac{24 \times 12}{4} = 72 \text{ in. controls}$$

$$b_w + 16t = 12 + 16(4) = 76 \text{ in.}$$

$$b' = 1/2[(8 \times 12) - 12] = 42$$

$$b = 2(42) + 12 = 96 \text{ in.}$$

Locate Neutral Axis

Using the bottom of the beam as a reference axis, take moments of areas.

		A	y	A_y
SLAB	$\dfrac{4 \times 72}{9}$ =	32	× 18 =	576
STEM	$\dfrac{12 \times 16}{9} - 11.8$ =	9.5	× 8 =	76.3
STEEL BEAM		11.8	× 10 =	118
		53.3		770.3

$$y_b = \frac{770.3}{53.3} = 14.5 \text{ in.}$$

Moment of Inertia of Composite Section

As in any reinforced concrete beam, neglect the concrete in the tensile zone below the neutral axis.

STEEL BEAM $\quad I_{CG} = 517$

$$Ad^2 = (11.8)(4.5)^2 = 239$$

SLAB $\quad I_{CG} = \dfrac{72 \times (4)^3}{12 \times 9} = 42.7$

$$Ad^2 = \left(\frac{72 \times 4}{9}\right)(3.5)^2 = 392$$

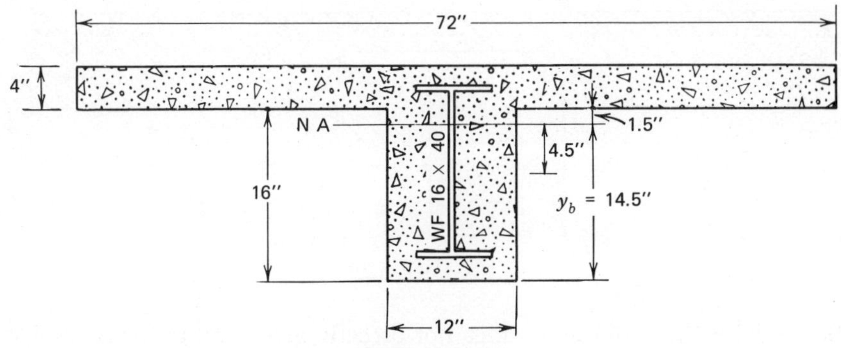

FIGURE 3.28

STEM $= \dfrac{12 \times (1.5)^3}{3 \times 9} \qquad = \dfrac{1.5}{1192.2 \text{ in.}^4}$

TENSILE S_{tr} $\dfrac{1192.2}{12.5} = 95.4 \text{ in.}^3$

COMPRESSIVE S_{tr} $\dfrac{1192.2}{5.5} = 216.8 \text{ in.}^3$

Check Section 1.11.2.2—AISC specifications:
Tensile S_{tr} shall not exceed

$$\left(1.35 + 0.35 \frac{Ml}{M_D}\right) S_s = \left[1.35 + 0.35\left(\frac{100.8}{45.2}\right)\right] 64.6$$

$$= 137.6 \text{ in.}^3 > 95.4 \text{ in.}^3 \quad \text{(no shoring required)}$$

Actual stresses in the composite section:

$$f_s = \frac{146 \times 12}{95.4} = 18.36 \text{ ksi} < 24 \text{ ksi} \qquad (0.66\ F_y)$$

$$f_c = \frac{146 \times 12}{216.8 \times 9} = 0.90 \text{ ksi} < 1.35 \text{ ksi} \qquad (0.45 f'_c)$$

Check Shear Connection

This type of beam has no shear connectors. The horizontal shearing force caused by bending the beam acts against the cross-hatched section shown in Figure 3.29.

However, the weakest plane in the section shown is along *a-b-c-d*. Therefore, the horizontal shear is carried by bond across line *b-c*, shear across lines *a-b* and *c-d*, and the mesh reinforcement that encases the

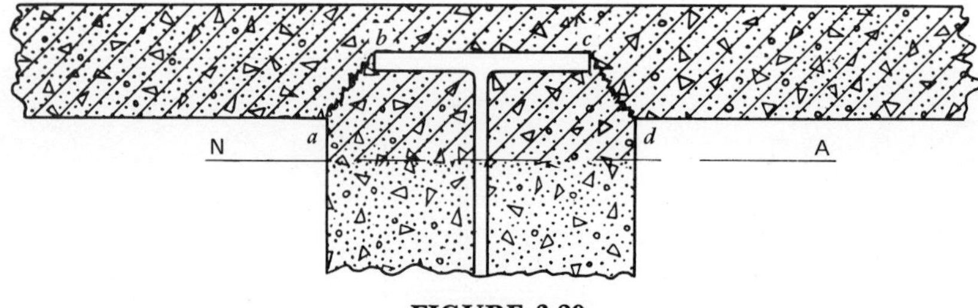

FIGURE 3.29

beam. The 1971 ACI code does not directly give a value for allowable bond stress. The value given in one of the older editions of the code, $0.03\,f'_c$, is a reasonable value to take for our case. The allowable shear stress across lines *a-b* and *c-d* is taken as $0.12\,f'_c$.

In calculating the applied shearing force, only the live load is used. For the applied shear, use the statics formula $v = VQ/I$. Value of Q of the concrete area above *a-b-c-d*:

It is sufficiently accurate and saves computation time to use the cross-hatched area shown in Figure 3.30.

$$Q = \frac{1}{9}[(4 \times 72 \times 3.5) - (2 \times 7 \times 2.5)] = 108\text{ in.}^3$$

$$V = 1400 \times \frac{24}{2} = 16{,}800\text{ lb}$$

$$v = \frac{(16{,}800)(108)}{1192.2} = 1522\text{ ppi}$$

Shearing Resistance

ALLOWABLE BOND STRESS	$0.03\,f'_c$	= 90 psi
ALLOWABLE SHEAR STRESS	$0.12\,f'_c$	= 360 psi
BOND ALONG *b-c*	(90)(7)	= 630 ppi
SHEAR ALONG *a-b* AND *c-d*	2(3.20)(360)	= 2304 ppi
		2934 ppi

2934 ppi > 1522 ppi

The shearing resistance is greater than the applied shear, so the beam is satisfactory.

Negative Moment Sections

The same section which was used for the positive moment will be used. Live and dead load moments will be two-thirds of the simple beam moments used in the previous calculations.

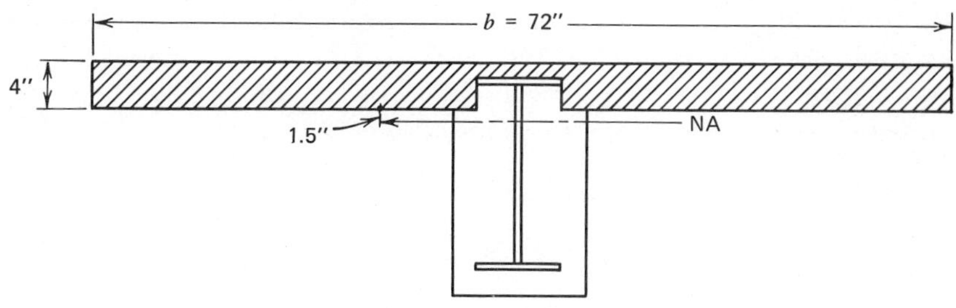

FIGURE 3.30

$$M_{DL} = 30.1 \text{ kip-ft}$$
$$M_{LL} = 67.2 \text{ kip-ft}$$
$$M_T = M_D + M_L = 97.3 \text{ kip-ft}$$

Referring to Figure 3.31, take moments of areas about axis A-A to locate the neutral axis. Ignore the concrete above the axis, which is now in tension.

$$\frac{12x}{9}\left(\frac{x}{2}\right) = 11.8(10 - x)$$

which yields

$$x = 7.13.$$

Moment of Inertia of Transformed Section

STEEL BEAM	I_{CG}	$= 517$
	$Ad^2 = 11.8(2.87)^2$	$= 97.2$
CONCRETE	$= \dfrac{12 \times (7.13)^3}{9 \times 3}$	$= \dfrac{161.1}{775.3 \text{ in.}^4}$

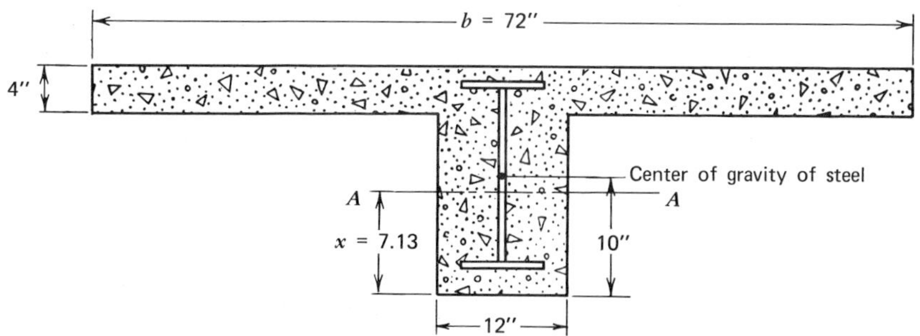

FIGURE 3.31 Encased beam–neutral axis for negative moment.

This simplified value of I is actually an approximation because it uses the entire compression area of the concrete, without subtracting out the space taken up by the steel heam. Rigorous methods have been developed for computing this moment of inertia, but the equations are clumsy and not suited to general office use. The approximate value of I is sufficiently accurate for design and construction purposes.

Computation of Stresses

DEAD LOAD ON THE STEEL BEAM

$$f_s = \frac{30.1(12)(8)}{517} = 5.6 \text{ ksi}$$

TOTAL LOAD ON THE COMPOSITE BEAM

$$f_s = \frac{(97.3)(12)(10.87)}{775.3} = 16.4 \text{ ksi} < 24 \text{ ksi}$$

$$f_c = \frac{(97.3)(12)(7.13)}{775.3 \times 9} = 1.19 \text{ ksi} < 1.35 \text{ ksi}$$

Shear in the Negative Moment Region

Resistance of the section to horizontal shear is a combination of shearing and bond strength, just as in the positive moment section of the beam. Since the bond resistance is much lower than the shear resistance, the critical section for horizontal shear is along the line of the bottom of the flange, as shown in Figure 3.32.

Use a live load shear, $V = 21{,}000$ lb.

$$Q = \left(\frac{12 \times 2}{9}\right)(7.13 - 1) = 16.3 \text{ in.}^3$$

$$v = \frac{VQ}{I} = \frac{(21{,}000)(16.3)}{775.3} = 442 \text{ ppi}$$

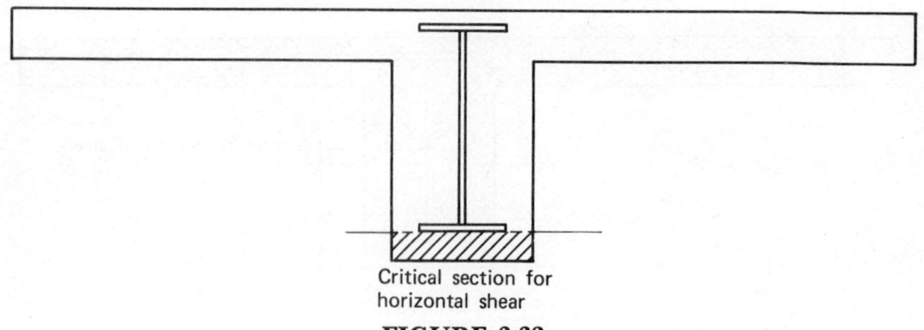

FIGURE 3.32

Shearing Resistance

BOND ALONG b-c	$7(90) =$	630 ppi
SHEAR ALONG a-b AND c-d	$2(2.5)(360) =$	1800 ppi
		2430 ppi

$$2430 \text{ ppi} > 442 \text{ ppi} \qquad \text{OK}$$

3.5 Concrete Slab on Wide Flange Beam (no coverplates by AISC specifications)

This beam is the standard composite beam. When the term composite construction is mentioned, this is the beam that naturally comes to mind. This type of composite beam has been the subject of more extensive research than any other kind of composite member.

This type of construction has been standardized so that the section can be designed by the usual transformed section method, or the section can be selected directly from the AISC composite beam tables. The examples that follow illustrate both of these methods and also check out the beam in the negative bending moment region.

Example 3.2

A-36 steel
Design of beam B-1

Loads
LIVE LOAD 150 psf
CEILING AND APPLIED FLOOR FINISH 25 psf
4-in. CONCRETE SLAB 50 psf

Allowable Stresses
$f'_c = 3000$ psi
$f_c = 1350$ psi
$n = 9$
$f_s = 24$ ksi

Compute the total bending moment and pick a trial section assuming the steel alone will carry all the loads. Since we can expect a 20 to 30% savings in steel with composite action, drop down two or three sizes and pick the steel beam. For example, if a W 16 × 50 section alone will carry the loads, drop down two sections and try the W 16 × 40 for the composite beam.

Loads Applied to the Steel Beam

SLAB	$8(50) =$	400 ppf
ASSUMED BEAM	$=$	40 ppf
		440 ppf

$$M_D = \frac{(0.440)(32)^2}{8} = 56.3 \text{ kip-ft}$$

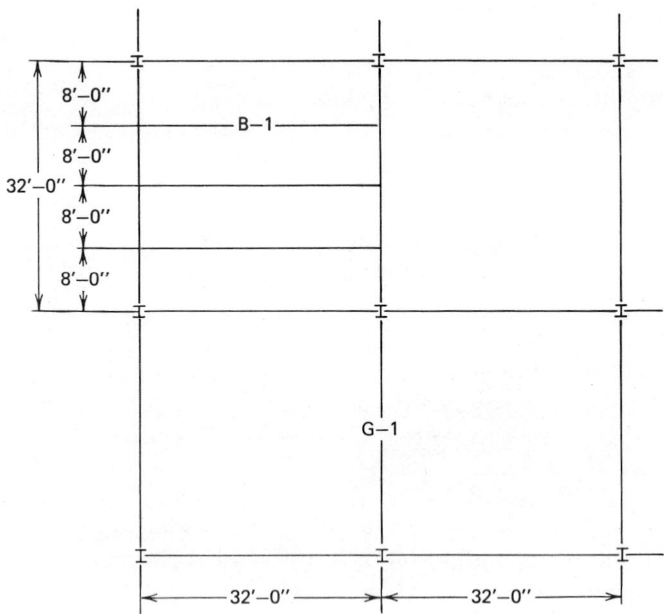

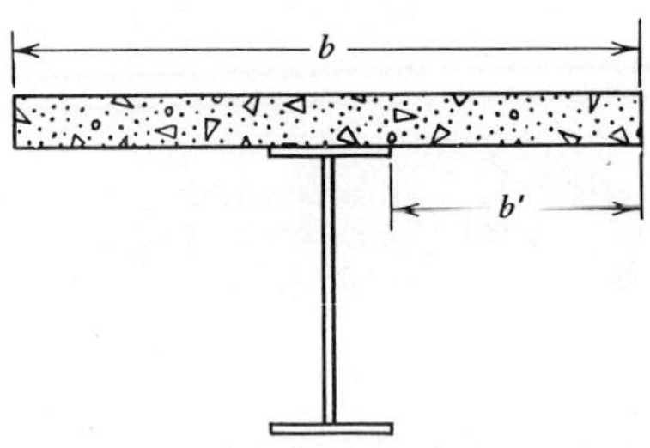

FIGURE 3.33 Beam layout.

Loads Applied to the Composite Beam

CEILING AND FLOOR FINISH	8(25) =	200 ppf
LIVE LOAD	8(150) =	1200 ppf
		1400 ppf

$$M_L = \frac{1.4(32)^2}{8} = 179.2 \text{ kip-ft}$$

$$M_T = M_D + M_L = 179.2 + 56.3 = 235.5 \text{ kip-ft}$$

Section Modulus Required

$$S = \frac{235.5(12)}{24} = 117.8 \text{ in.}^3$$

A wide flange section, W 18 × 64, furnishes 118 in.³ of section modulus, so drop down three sections and try the W 18 × 50 for the composite beam.

$$S = 89.1 \text{ in.}^3$$
$$I = 802 \text{ in.}^4$$
$$\text{Area} = 14.7 \text{ in.}^2$$
$$\text{Depth} = 18.0 \text{ in.}$$
$$\text{Flange width, } b_f = 7.5 \text{ in.}$$

Check effective flange width of composite section.

$$b \leqslant 1/4 \text{ span} = \frac{32 \times 12}{4} = 96 \text{ in.}$$
$$b \leqslant c/c \text{ beams} = 8(12) = 96 \text{ in.}$$
$$b \leqslant 16t + b_f = 16(4) + 7.5 = 71.5 \text{ in.} \longleftarrow \text{controls}$$

Properties of the Composite Section

Locate the neutral axis:

	A	y	A_y
SLAB	$\frac{4 \times 71.5}{9} = 31.8$	× 20 =	635.6
BEAM	14.7	× 9 =	132.3
	46.5		767.9

$$\bar{y} = \frac{767.9}{46.5} = 16.5 \text{ in.}$$

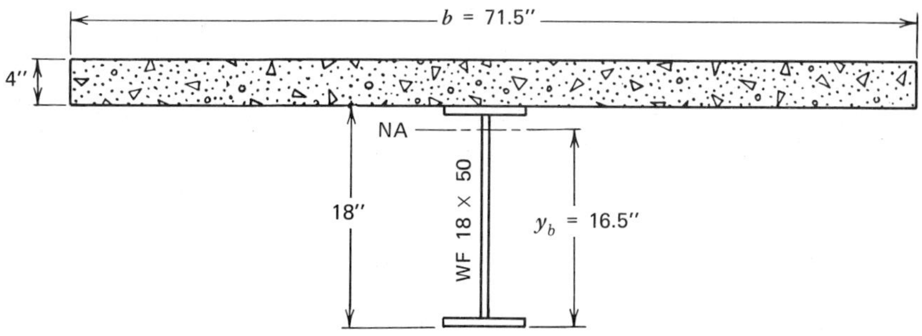

FIGURE 3.34 Section dimensions.

I *of Transformed Section*

SLAB $$I_{CG} = \frac{71.5 \times (4)^3}{12(9)} = 42.4 \text{ in.}^4$$

$$Ad^2 = \left(\frac{71.5 \times 4}{9}\right)(3.5)^2 = 389.3$$

STEEL BEAM $$I_{CG} = \qquad = 802.0$$

$$Ad^2 = 14.7(7.5)^2 = 826.9$$

$$2060 \text{ in.}^4$$

TENSILE S_{tr} $$\frac{2060}{16.5} = 124.8 \text{ in.}^3$$

COMPRESSIVE S_{tr} $$\frac{2060}{5.5} = 374.5 \text{ in.}^3$$

Check AISC Formula 1.11.2.

$$S_{tr} = \left(1.35 + 0.35 \frac{M_L}{M_D}\right) S_s$$

$$= \left[1.35 + 0.35 \left(\frac{179.2}{56.3}\right)\right] 89.1 = 219.5 \text{ in.}^3$$

$$219.5 \text{ in.}^3 > 124.8 \text{ in.}^3 \qquad \text{OK}$$

Stress carried by the steel beam alone:

$$f_s = \frac{M_D}{S_s} = \frac{56.3(12)}{89.1} = 7.58 \text{ ksi}$$

Now check the construction conditions to determine the allowable stress in the compression flange of the steel section. AISC Formula 1.5-7 yields an allowable stress as follows:

$$F_b = \frac{12 \times 10^3 \, C_b}{(1d/A_f)} = \frac{12{,}000}{(52 \times 12)(4.21)} = 7.42 \text{ ksi}$$

The dead load bending stress exceeds the allowable value. There are two logical ways to increase the allowable stress value. If the beam were designed as continuous, the unsupported length of compression flange would be the distance between points of zero moment. This would drop the unsupported length from 32 ft to about 20 ft and raise the allowable stress to about 12 ksi.

The other method of raising the allowable stress would be to provide intermediate bracing for the compression flange until the concrete had gained at least 75% of its strength.

Stresses on the composite section after the concrete hardens:

$$f_s = \frac{235.5(12)}{124.8} = 22.6 \text{ ksi} < 24 \text{ ksi} \qquad \text{OK}$$

$$f_c = \frac{235.5(12)}{374.5(9)} = 0.84 \text{ ksi} < 1.35 \text{ ksi} \qquad \text{OK}$$

Shear connectors:

$$V_h = \frac{0.85\, f'_c A_c}{2} = \frac{(0.85)(3)(4 \times 71.5)}{2} = 365 \text{ kips}$$

$$V_h = \frac{A_s F_y}{2} = \frac{(14.7)(36)}{2} = 265 \text{ kips} \longleftarrow \text{controls}$$

Use the $\frac{3}{4} \times 3$-in. stud. From AISC capacity tables, this stud carries 11.5 kips.

Number of connectors $= 265/11.5 = 23$ connectors on each side of the beam centerline (C_L) required. In this case, 24 on each side would be a better choice because this gives a uniform spacing of $(32 \times 12)/(2 \times 24) =$ 8 in., which makes detailing and construction easier.

The values in this example agree with those given in the AISC composite beam tables for the W 18×50 with a 4-in. slab.

The efficiency of the coverplated beam can be seen in these same tables. A steel beam W 16×26 with a $1 \times 4\frac{1}{2}$ in. coverplate furnishes the same capacity. This coverplated beam weighs 41 lb/ft, which saves an additional 18% of steel weight. It is true that a large part of the cost of a plated beam is the labor cost. However, part of this can be recovered by eliminating the plate in the low stress regions where it is not required.

Check the same W 18×50 beam in the negative moment region. Note that this section was designed as a simply supported beam, using $M = \frac{1}{8} wL^2$ as the maximum bending moment. If the beam had initially been considered as continuous, both the positive and negative moments would be less than this value.

Assume

$$M_D = 50 \text{ kip-ft}$$

$$M_L = 162 \text{ kip-ft}$$

$$M_T = 162 + 50 = 212 \text{ kip-ft}$$

The slab is now in the tension zone, and the bottom flange of the steel beam is the compressive region. Since the slab is cracked, the tension will be taken by the reinforcing bars in the slab. Assume No. 8 bars at 6-in. spacing. A_s for the bars is $(1.57 \text{ in.}^2/\text{ft})(71.5/12) = 9.34 \text{ in.}^2$ of steel.

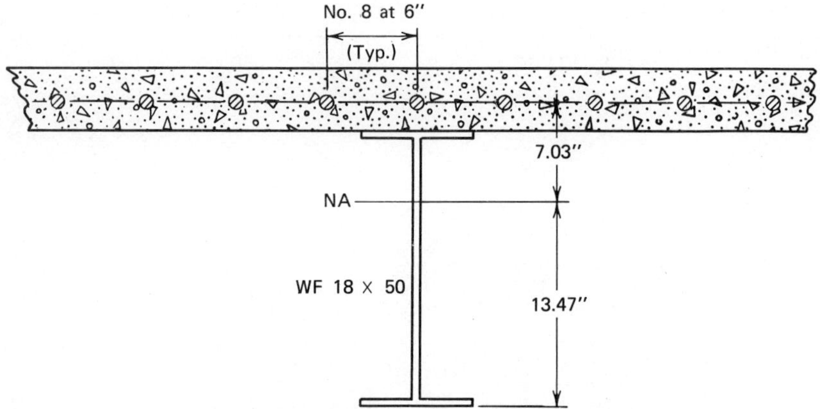

FIGURE 3.35 Negative moment section.

Properties of the Composite Section

Locate the neutral axis:

	A	y	A_y
BEAM	14.7 ×	9 =	132.3
SLAB BARS	9.34 ×	20.5 =	191.5
	24.04		323.8

$$\bar{y} = \frac{323.8}{24.04} = 13.47 \text{ in.}$$

Moment of Inertia

BEAM I_o		802 in.^{4}
BEAM Ad^2	$14.7(4.47)^2 =$	294
REINFORCING BARS Ad^2	$(9.34)(7.03)^2 =$	462
		1558 in.^{4}

TENSILE S_{tr} $\quad \dfrac{1558}{7.03} = 221 \text{ in.}^3$

COMPRESSIVE S_{tr} $\quad \dfrac{1588}{13.47} = 115.6 \text{ in.}^3$

Stress in steel beam (DL only):

$$f_s = \frac{M_D}{S_s} = \frac{50(12)}{13.47} = 6.73 \text{ ksi}$$

Stress in composite beam:

BEAM f_s $\dfrac{(212)(12)}{115.6} = 22 \text{ ksi} < 24 \text{ ksi}$ OK

BARS f_s $\dfrac{(212)(12)}{221} = 11.5 \text{ ksi} < 24 \text{ ksi}$ OK

In continuous composite beams, it is important to supply shear connectors in the negative moment region, as well as the positive moment portions of the beam. In the negative moment region, enough shear connectors are required to transfer from the slab reinforcement to the steel beam, one-half of the ultimate tensile strength of the reinforcement.

$$V_h = \frac{A_{sr}F_{yr}}{2}$$

where A_{sr} and F_{yr} are the values for the reinforcing steel contained in the effective flange width,

$$V_h = \frac{(9.34)(60)}{2} = 280.2 \text{ kips}$$

Using the $3/4 \times 3$-in. headed stud, which has a capacity of 11.5 kips, the number of connectors required is

$$N = \frac{280.2}{11.5} = 24 \text{ connectors}$$

For full composite action, these connectors would have to be placed on either side of the point of maximum negative moment, from the point of maximum moment to the point of zero moment. If our 32-ft span were one bay of a building three bays wide, the negative moment region is defined approximately in Figure 3.36. Since the negative moment region

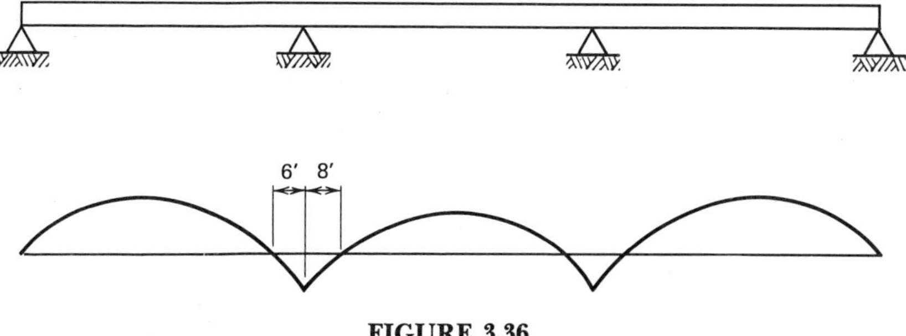

FIGURE 3.36

is unsymmetrical with respect to the support, the connector spacing can vary on the two sides of the support. A different spacing is also required in the positive moment region. Both detailing and construction supervision will be simplified by using only two spacings, one in the positive region and one in the negative region, even though it uses a few extra studs. However, if this beam were repeated many times in the overall building, the detailing of three stud spacings for the beam could be economically justified.

Reviewing the entire beam, the stud spacings would be as follows:

POSITIVE REGION SPAN 1 AND 3 $$-\frac{26 \times 12}{2 \times 24} = 6.5 \text{ in.}$$

NEGATIVE REGION SPAN 1 AND 3 $$\frac{6 \times 12}{24} = 3 \text{ in.}$$

NEGATIVE REGION SPAN 2 $$\frac{8 \times 12}{24} = 4 \text{ in.}$$

POSITIVE REGION SPAN 2 $$\frac{16 \times 12}{2 \times 24} = 4 \text{ in.}$$

For simpler detailing and construction, the actual spacings shown in Figure 3.37 could be used. This assumes, however, that full composite action is desired. The use of partial composite action would spread out these stud spacings and make the beam more practical and economical.

These spacings all assume that the connectors are placed singly and symmetrically over the beam web, as shown in Figure 3.38*a*. The spacings could be doubled by placing the connectors in pairs, as shown in Figure 3.38*b*. The limitation on placing the studs in pairs is that the studs tend to tear out of thin flanges before developing their full strength. Quantitatively, the limitation is that the stud diameter shall not be greater than 2.5 times the flange thickness. In our case, with $\frac{3}{4}$-in. studs and a flange thickness slightly greater than $\frac{1}{2}$ in., the studs can easily be placed in pairs and the spacing doubled.

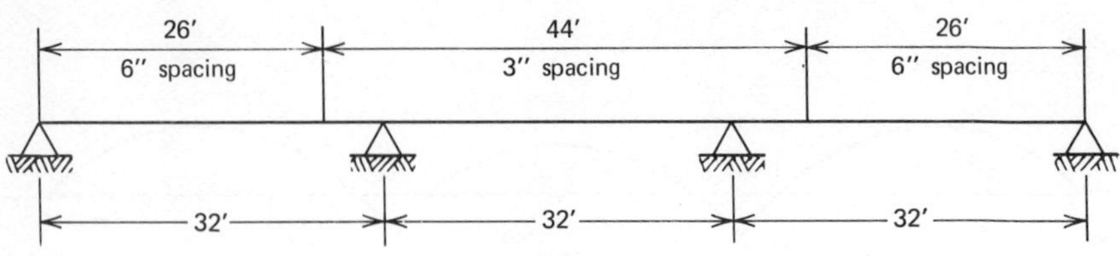

FIGURE 3.37

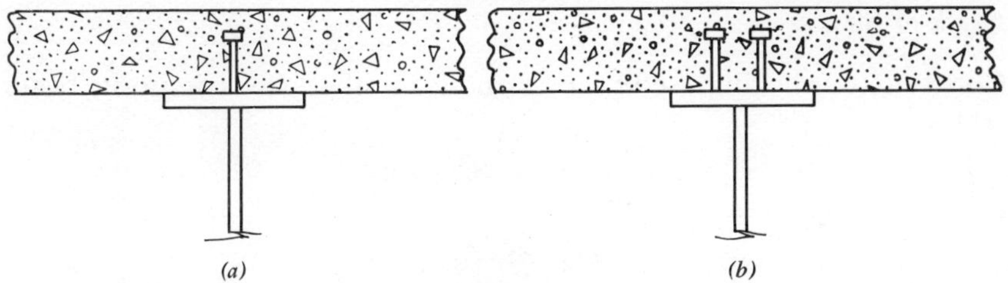

(a) (b)

FIGURE 3.38

Check the previous example using the composite beam tables in the AISC manual for 36-ksi yield point steel.

$$M_D = 56.3 \text{ kip-ft}$$
$$M_L = 179.2 \text{ kip-ft}$$
$$M_T = M_D + M_L = 235.5 \text{ kip-ft}$$

Maximum vertical shear:

$$V = (0.44 + 1.40)\frac{32}{2} = 29.44 \text{ kips}$$

Effective width, $b = 71.5$ in.

Required Section Moduli:

For M_T,

$$S_{tr} = \frac{235.5 \times 12}{24} = 117.8 \text{ in.}^3$$

For M_D,

$$S_s = \frac{56.3 \times 12}{24} = 28.2 \text{ in.}^3$$

Use properties table for a 4-in. slab with $S_{tr} = 117.8$ (no coverplate). Select W 18×50.

$$b = 71.5 \text{ in.}$$
$$y_b = 16.52 \text{ in.}$$
$$W_s = 50 \text{ ppf}$$
$$V = 93 \text{ kips}$$
$$S_{tr} = 125 \text{ in.}^3$$
$$S_t = 376 \text{ in.}^3$$
$$S_s = 89.1 \text{ in.}^3$$

Check of Stresses

Concrete (no shoring used); see footnote 2 in the AISC composite beam tables.

$$\frac{M_L}{M_D} = 3.18 \qquad \text{use } 3.0$$

$$\text{Max } \frac{S_{tr}}{S_t} = 0.67$$

$$\text{Actual } \frac{S_{tr}}{S_t} = \frac{125}{376} = 0.33 < 0.67 \qquad \text{concrete stress} \qquad \text{OK}$$

Section furnished		*Section required*	
STEEL TOTAL LOAD	125 in.^{3}	117.8 in.^{3}	OK
DEAD LOAD	89.1 in.^{3}	28.2 in.^{3}	OK
WEB SHEAR	93 kips	29.4 kips	OK

Deflection Check

$$\frac{L}{360} = \frac{32 \times 12}{360} = 1.06 \text{ in.} \qquad \text{allowable for live load}$$

$$\text{DL} = \frac{M_D L^2}{160 S_s y_{bs}} = \frac{56.3 \times (32)^2}{160(39.1)(9.0)} = 0.45 \text{ in.}$$

$$\text{LL} = \frac{M_L L^2}{160 S_{tr} y_b} = \frac{179.2(32)^2}{160(125)(16.52)} = 0.55 < 1.06 \qquad \text{OK}$$

Check AISC Formula 1.11.2.

$$S_{tr} = \left(1.35 + 0.35 \frac{M_L}{M_D}\right) S_s$$

$$= [1.35 + 0.35(3.18)]89.1 = 219.4 \text{ in.}^3 > 125 \text{ in.}^3 \qquad \text{OK}$$

Shear connectors (full composite action):
Use $\frac{3}{4} \times 3$-in. studs. The $\frac{3}{4}$-in. size is less than $2\frac{1}{2}$ times the flange thickness.

"Stud Coefficient" Method

$$N_s = W_s \times \text{stud coefficient}$$
$$= 50 \times 0.461 = 23.05$$

$$N_c = Ac \times \text{stud coefficient}$$
$$= (4 \times 71.5) \times 0.111 = 31.75$$

Use 24 connectors on each side of the point of maximum moment for a total of 48 studs.

Example 3.3 Coverplated Beam

The coverplated beam makes an efficient composite beam because most of the steel is in the bottom flange where it is needed. The top flange, which is close to the neutral axis, is much smaller. However, this small top flange poses a real construction problem. Either the beam must be shored, or else positive lateral support must be provided for this relatively weak compression flange until the concrete cures.

Redesign Example 3.2 using a coverplated beam and AISC specifications.

SPAN 32 ft

EFFECTIVE FLANGE WIDTH 71.5 in.

$$M_D = 56.3 \text{ kip-ft}$$

$$M_T = M_D + M_L = 235.5 \text{ kip-ft}$$

BEAM SPACING 8 ft

$$M_L = 179.2 \text{ kip-ft}$$

Composite Section Modulus Required:

$$S_{tr} = \frac{235.5(12)}{24} = 117.8 \text{ in.}^3$$

Steel Section Modulus Required

1. If full lateral support of the compression flange is assured,

$$S_s = \frac{56.3(12)}{24} = 117.8 \text{ in.}^3$$

2. If lateral support is not provided, the allowable stress is not known until the section properties are chosen. Select a section based on full lateral support, and then check the allowable stress in the compression flange of the steel section.

Interpolating from the AISC composite beam tables, try W 16×26 with 1×4-in. plate and 4-in.-thick slab.

Steel Section Properties

		A	y	A_y
PLATE	$1 \times 4 =$	4.0	$\times 0.5 =$	2.0
BEAM		7.67	$\times 8.83 =$	67.7
		11.67		69.7

$$y_{bs} = \frac{69.7}{11.67} = 5.97 \text{ in.}$$

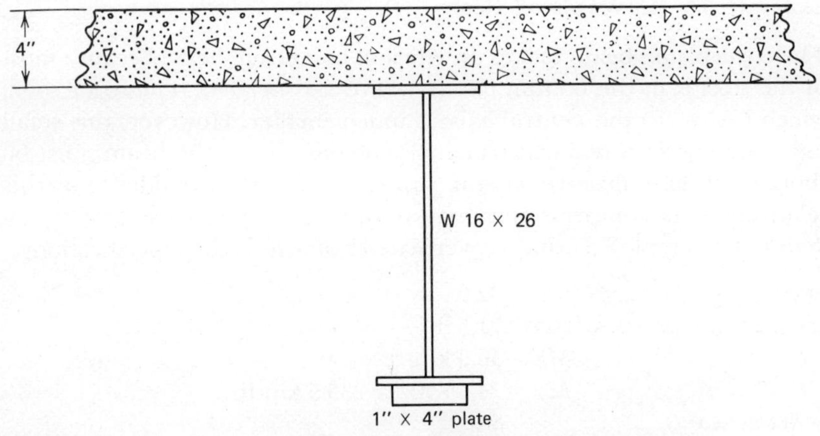

FIGURE 3.39 Coverplated beam.

I *of Steel Section*

BEAM	I_{CG}	$= 300$ in.^{4}
	$Ad^2 = (7.67)(2.86)^2 =$	62.7
PLATE	$Ad^2 = (4)(5.47)^2 =$	119.7
		482.4 in.^{4}

SECTION MODULUS, TOP $\dfrac{482.4}{10.68} = 45.2 \text{ in.}^3$

SECTION MODULUS, BOTTOM $\dfrac{482.4}{5.97} = 80.8 \text{ in.}^3$

Composite Section Properties

	A		y		A_y
PLATE	4	×	0.5	=	2.0
BEAM	7.67	×	8.83	=	67.7
SLAB $\dfrac{4 \times 71.5}{9} =$	31.78	×	18.65	=	592.66
	43.45				662.35

$$y_b = \frac{662.35}{43.45} = 15.24 \text{ in.}$$

I *of Composite Section*

BEAM	$I_{CG} =$	$=$ 300 in.^{4}
	$Ad^2 = (7.67)(6.41)^2$	$=$ 315 in.^{4}
PLATE	$Ad^2 = (4)(14.74)^2$	$=$ 869 in.^{4}

SLAB $I_{CG} = \frac{(71.5)(4)^3}{(12)(9)} = 42 \text{ in.}^4$

$$Ad^2 = \left(\frac{71.5 \times 4}{9}\right)(3.41)^2 = \underline{370 \text{ in.}^4}$$

$$1869 \text{ in.}^4$$

TENSILE S_{tr} $\frac{1869}{15.24} = 124.4 \text{ in.}^3 > 117.8 \text{ in.}^3$

COMPRESSIVE S_{tr} $\frac{1869}{5.41} = 363.8 \text{ in.}^3$

Check AISC Formula 1.11.2.

$$S_{tr} = \left(1.35 + 0.35\frac{M_L}{M_D}\right)S_s$$

$$= \left[1.35 + 0.35\left(\frac{179.2}{56.3}\right)\right]80.8 = 199 \text{ in.}^3 > 124.4 \text{ in.}^3 \qquad \text{OK}$$

Tensile stress carried by the steel beam alone:

$$f_s = \frac{M_D}{S_s} = \frac{(56.3)(12)}{80.8} = 8.36 \text{ ksi} \qquad \text{OK}$$

Stresses on the composite section after the concrete hardens:

$$f_s = \frac{235.5(12)}{124.4} = 22.7 \text{ ksi} \qquad \text{OK}$$

$$f_c = \frac{235.5(12)}{363.8(9)} = 0.86 \text{ ksi} < 1.35 \text{ ksi} \qquad \text{OK}$$

The composite section is apparently satisfactory. However, now check the steel section, assuming there is no lateral support for the compression flange. From AISC Formula 1.5–6b, for the W 16×26 section, $r_T = 1.38$:

$$\frac{L}{r_T} = \frac{32 \times 12}{1.38} = 278$$

$$F_b = \frac{170 \times 10^3}{(L/r_T)^2} = \frac{170{,}000}{(278)^2} = 2.2 \text{ ksi}$$

With our section, the section modulus related to the top flange is 45.2 in.^{3}

TOP FLANGE STRESS $\frac{(56.3)(12)}{45.2} = 15 \text{ ksi}$

The top flange stress at 15/2.2 is almost seven times the allowable value. This illustrates the importance of construction method. The formwork

for this beam must be designed to provide lateral support or the beam will probably fail.
Shear connectors:

$$V_h = \frac{0.85f'_cA_c}{2} = \frac{(0.85)(3.0)(4\times 71.5)}{2} = 365 \text{ kips}$$

$$V_h = \frac{A_sF_y}{2} = \frac{(11.67)(36)}{2} = 210 \text{ kips} \qquad \text{(governs)}$$

Using 3-in.-long sections of C 3×4.1 channel, each connector can resist $3\times4.3 = 12.9$ kips. The number required $= 210/12.9 = 16.3$. Use 17 connectors on each side of the centerline, equally spaced.
Coverplate cutoff:

RESISTING MOMENT, PLATED SECTION $= \dfrac{24(124.4)}{12} = 248.8$ kip-ft

RESISTING MOMENT, UNPLATED SECTION $= \dfrac{24(80.8)}{12} = 161.6$ kip-ft

$$\frac{x^2}{(L/2)^2} = \frac{(248.8-161.6)}{248.8}$$

$$x = \frac{87.2}{248.8}(16)^2 \qquad x = 9.5 \text{ ft}$$

Using E70xx electrodes and $\frac{5}{16}$-in. welds, the plates are extended (AISC Article 1.10.4) $1\frac{1}{2}$ times the plate width past the theoretical cutoff point, with continuous weld along both sides and the end of the plate.
TOTAL LENGTH OF THE PLATE $= 2(9.5\text{ ft}) + 2(6\text{ in.}) = 20$ ft

Example 3.4 Beam with Flange on One Side Only

The beam with partial slab on one side only is usually the exterior or perimeter beam in the building. This type of composite beam needs special care in construction because the loads are mostly on one side of the beam which tends to twist the beam out of its plane. The beam must be braced during construction to counteract this torsional effect.

In modern curtain-wall construction, the exterior sheathing of the building is often connected to the exterior beam. However, this load is not applied until after the slab has cured and the beam has become composite. For Example 3.4, a precast sheathing panel supported by the exterior beam is used. The loading is shown in Figure 3.40.

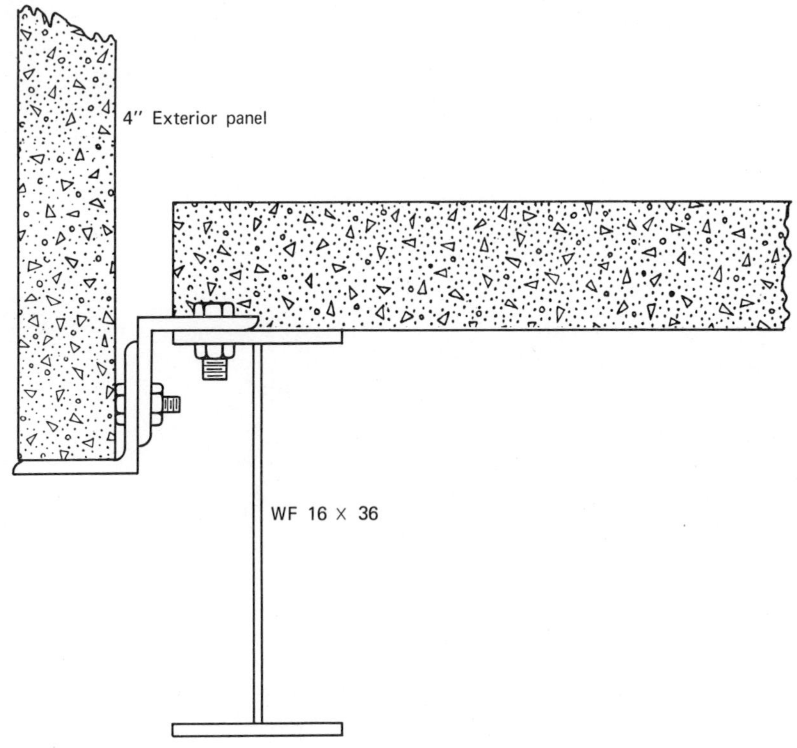

FIGURE 4.40 Exterior beam.

Loads
LIVE LOAD 100 psf
4-in. CONCRETE SLAB 50 psf
CEILING AND FLOOR FINISH 25 psf
INTERIOR WALL 100 ppf
EXTERIOR SHEATHING 600 ppf

Allowable Stresses
$f'_c = 3{,}000$ psi
$f_c = 1{,}350$ psi
$n = 9$
A-36 steel
$f_s = 24{,}000$ psi

SPAN 30 ft
BEAM SPACING 7.5 ft

Maximum vertical shear applied:

$$V = (0.218 + 1.169)\frac{30}{2} = 20.8 \text{ kips}$$

DEAD LOAD—SLAB $50 \times 3.75 = 187.5$ ppf
Est. Beam $= 30$ ppf
218.5 ppf

Loads on Composite Beam

INTERIOR WALL = 100 ppf
EXTERIOR WALL = 600 ppf
CEILING AND FLOOR = 25 × 3.75 = 94 ppf
LIVE LOAD = 100 × 3.75 = 375 ppf
1169 ppf

$$M_D = \frac{0.218(30)^2}{8} = 24.5 \text{ kip-ft}$$

$$M_L = \frac{1.169(30)^2}{8} = 131.5 \text{ kip-ft}$$

$$M_T = M_D + M_L = 156.0 \text{ kip-ft}$$

$$\text{ESTIMATED } S_{tr} = \frac{156.0(12)}{24} = 78.0 \text{ in.}^3$$

Use the composite beam tables for a partial slab 4 in. thick and $S_{tr} = 78.0$ (try W 16 × 36).

Section properties from the table:

$$S_{tr} = 78.0 \text{ in.}^3$$
$$S_{tr} = 167 \text{ in.}^3$$
$$S_s = 56.5 \text{ in.}^3$$
$$I_{tr} = 1060 \text{ in.}^4$$
$$y_b = 13.53 \text{ in.}$$
$$b = 31 \text{ in.}$$

Check effective width of slab:

$$b = 6t + b_f \qquad = 6(4) + 7 = 31 \text{ in.} \longleftarrow \text{governs}$$

$$b = \frac{1}{12}\text{span} + b_f \qquad = \frac{30(12)}{12} + 7 = 37 \text{ in.}$$

$$b = \tfrac{1}{2}\text{clear distance} + b_f = 41.5 + 7 \qquad = 48.5 \text{ in.}$$

Tensile stress carried by the steel beam alone:

$$f_s = \frac{M_D}{S_s} = \frac{24.5(12)}{56.5} = 5.2 \text{ ksi}$$

Stresses carried by the composite section:

$$f_s = \frac{(156)(12)}{78} = 24 \text{ ksi} \qquad \text{OK}$$

$$f_c = \frac{156(12)}{167(9)} = 1.25 \text{ ksi} \qquad \text{OK}$$

The composite section looks satisfactory. However, the loads applied to the steel beam before the concrete hardens are eccentric loads, so that the resistance of the steel beam to lateral buckling of the compression flange becomes especially important.

AISC Formula 1.5–7:

$$\text{W } 16 \times 36 \qquad \frac{d}{A_f} = 5.30$$

$$F_b = \frac{12 \times 10^3}{(30 \times 12)(5.30)} = 6.3 \text{ ksi} > 5.2 \text{ ksi} \qquad \text{OK}$$

Vertical shear is seldom a factor in composite design. In general, short-span, heavily loaded beams tend to be critical in shear, and long spans tend to be critical in bending. Composite construction is used most often with longer spans. However, check the vertical shear for this particular example:

APPLIED SHEAR (from above) $= 20.8$ kips

Shear Resistance of the Section

$$V = \text{web area} \times \text{allowable stress}$$

$$v = (15.85)(0.299)(14.4) = 68.2 \text{ kips} \qquad \text{OK}$$

Shear Connectors—Stud Coefficient Method

This method, from AISC, gives two values. One value is based on the weight of steel and the other is based on concrete area. Coefficients are given in the AISC manual. Use $\frac{5}{8}$ in. $\times$ $2\frac{1}{2}$ in. studs.

$$N_s = 0.662 \times 36 = 23.8$$

$$N_c = 0.160 \times 4 \times 31 = 19.8 \longleftarrow \text{Governs}$$

Use 20 studs on either side of the beam centerline, or 40 studs in all.

Example 3.5 Strengthen an Existing Girder by Adding Shear Connectors and a New Slab—AASHTO Specifications

Composite construction is sometimes used to increase the carrying capacity of an existing structure. In the following example, a plate girder bridge is modified by removing the old slab, adding shear connectors, and then placing a new slab. The new plate girder was designed by AASHTO specifications for HS-20 loading. The symmetrical noncomposite girder is shown in Figure 3.41.

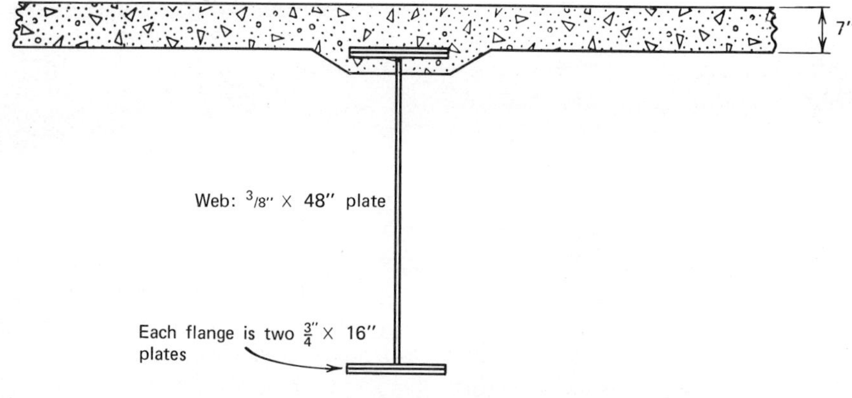

FIGURE 3.41

Existing Girder

WEB	$\frac{3}{8} \times 48$ in.
FLANGES	two $\frac{3}{4} \times 16$ in. plates
SLAB THICKNESS	7 in.
SPAN	80 ft

Girder Capacity Moment of Inertia

WEB $$\frac{3}{8} \times \frac{(48)^3}{12} = 3{,}456 \text{ in.}^4$$

FLANGES $$2(1.5 \times 16)(24.75)^2 = \underline{29{,}403 \text{ in.}^4}$$
$$32{,}859 \text{ in.}^4$$

RESISTING M $$\frac{20(32{,}859)}{12 \times 25.5} = 2{,}148 \text{ kip-ft}$$

Now check the resisting capacity of the same girder with a 7-in. slab bonded by shear connectors.

EFFECTIVE FLANGE WIDTH $\quad 12t = 7 \times 12 = 84$ in.

Figure 3.42 shows the composite girder.

Moment of Inertia of Composite Section

AREA OF STEEL SECTION $$2(1.5 \times 16) + \left(\frac{3}{8} \times 48\right) = 66 \text{ in.}^2$$

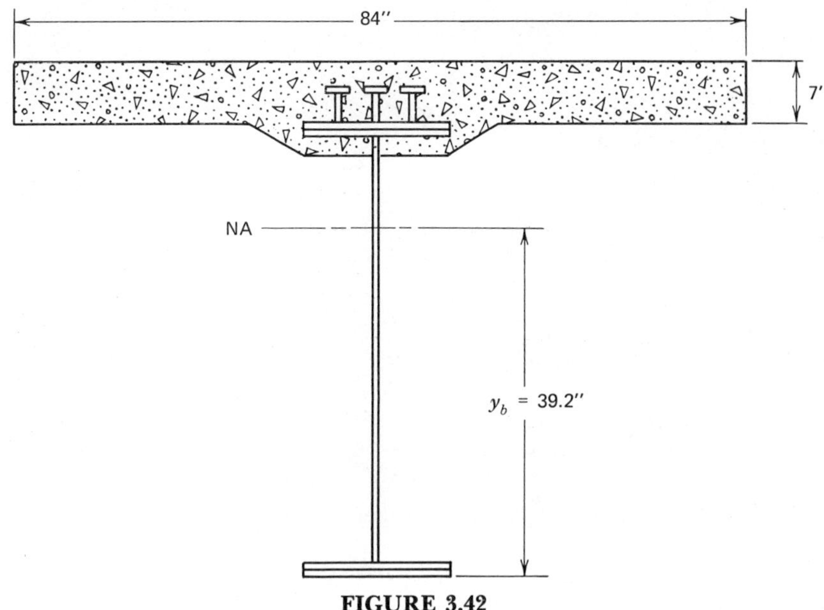

FIGURE 3.42

TRANSFORMED SLAB AREA $\dfrac{7 \times 84}{10} = 58.8 \text{ in.}^2$

$$\text{Total area} = 124.8 \text{ in.}^2$$

$$y_b = \frac{1}{124.8}\left[\left(66 \times \frac{51}{2}\right) + (58.8 \times 54.5)\right] = 39.2 \text{ in.}$$

Moment of Inertia

BOTTOM PLATES	$(1.5 \times 16)(39.2 - 0.75)^2$	$= 35{,}481$
WEB I_o	$\dfrac{3}{8} \times \dfrac{(48)^3}{21}$	$= 3{,}456$
WEB Ad^2	$\left(\dfrac{3}{8} \times 48\right)(13.7)^2$	$= 3{,}378$
TOP PLATES	$(1.5 \times 16)(11.05)^2$	$= 2{,}930$
SLAB	$(58.8)(15.3)^2$	$= 13{,}764$
		$59{,}009 \text{ in.}^4$

Resisting Moments

BOTTOM FLANGE $\dfrac{20(59{,}009)}{12(39.2)} = 2{,}509 \text{ kip-ft}$

TOP FLANGE $\dfrac{1.35(59{,}009)(10)}{12(18.8)} = 3{,}531$ kip-ft

Increase in capacity is 17%.

Shear Connectors by AASHTO

The AASHTO specifications state that shear connectors shall be designed for fatigue and then checked for ultimate strength. The resisting capacity of a stud connector, Z_r, for fatigue is given by the following:

$$Z_r = \alpha d^2$$

where Z_r = allowable capacity of a shear connector (lb);
H = height of stud (in.);
d = diameter of stud (in.);
α = a fatigue constant, 13,000 for 100,000 cycles, 10,600 for 500,000 cycles, and 7,850 for 2,000,000 cycles.

For the headed stud connectors, the ratio of (H/d) should be equal to or greater than 4.

For the design, assume $V = 125$ kips. The applied shear is computed from the statics formula, $v = VQ/I$. Use $3\frac{1}{2} \times \frac{7}{8}$ headed studs $= H/d = 4.0$.

$$Q = \left(\frac{7 \times 84}{10}\right)(15.3) = 900 \text{ in.}^3$$

$$v \text{ AT SUPPORT} = \frac{125(900)}{59{,}009} = 1.9 \text{ kip/in.}$$

$$v \text{ AT 1/4 POINT} = \frac{62.5(900)}{59{,}009} = 0.95 \text{ kip/in.}$$

Assuming 2,000,000 cycles, $\alpha = 7850$.

CAPACITY OF ONE STUD $7850(7/8)^2 = 6010 \text{ lb} = 6.01$ kips.

Using three studs at each section:

SPACING AT SUPPORT $\dfrac{3 \times 6.01}{1.9} = 9.5$ in.

SPACING AT 1/4 POINT $\dfrac{3 \times 6.01}{0.95} = 19$ in.

At this point, it is probably best to solve for the spacings at several sections along the beam and to lay out a feasible spacing diagram to determine the total number of connectors necessary to satisfy the fatigue requirement. If this number of connectors is greater than the number required by the ultimate strength requirement, the greater number shall be used. However, AASHTO does permit the total number of connectors to be spaced uniformly.

For our 80 ft span, a connector spacing that would satisfy the fatigue requirement follows:

1 at 6 in. from support
6 at 9 in.
17 at 12 in.
12 at 18 in.

This gives $36 \times 3 = 108$ connectors on each side of the beam centerline. By solving for the shear at more frequent intervals, a more economical spacing can be worked out.

Ultimate Strength Check

CAPACITY OF ONE STUD, $S_u = 930d^2\sqrt{f'_c}$

$$= 930\left(\frac{7}{8}\right)^2 \sqrt{3000} = 39 \text{ kips}$$

$$P_1 = A_sF_y = (66)(30) = 2376 \text{ kips}$$

$$P_2 = 0.85f'_cbt = (0.85)(3.0)(84 \times 7)$$

$$= 1500 \text{ kips} \longleftarrow \text{controls}$$

$$N = \frac{P}{\phi S_u} = \frac{1500}{(0.85)(39)} = 45 \text{ connectors each side of the beam centerline}$$

Use the fatigue requirement: 36 rows of three studs on each side of beam centerline. Place the first row of studs 2 in. in from the support, and then use a uniform spacing of 12 in.

In this particular case, construction cost would make this bridge a poor candidate for strengthening. Removing the slab, installing the shear connectors, and casting the new slab only increase the carrying capacity of the bridge by 17%. If the existing slab were badly deteriorated so that the bridge was already a candidate for rehabilitation work, the strengthening by composite action would be reasonable.

3.6 The Box Girder

The box girder is actually a plate girder with two webs, fabricated in the form of a hollow box. The box girder furnishes exceptionally good torsional resistance because of its shape. One of the more beautiful composite box girder bridges in this country is the Latah Creek Bridge, which is shown as the frontispiece of this book.

Example 3.6 Composite Box Girder

This is the main girder G-1 shown in the floor plan in Figure 3.43. The loading is taken as follows:

DEAD LOAD (concentration from B-1) $= 21.75$ kips
UNIFORM LOAD $= 0.525$ kpf

$$M_D = 21.75(10) + \frac{0.525(30)^2}{8}$$

$$= 276.6 \text{ kip-ft}$$

LIVE LOAD (concentration from B-1) $= 43.75$ kips
UNIFORM LOAD $= 0.875$ kpf

$$M_L = 43.75(10) + \frac{0.875(30)^2}{8}$$

$$= 535.9 \text{ kip-ft}$$

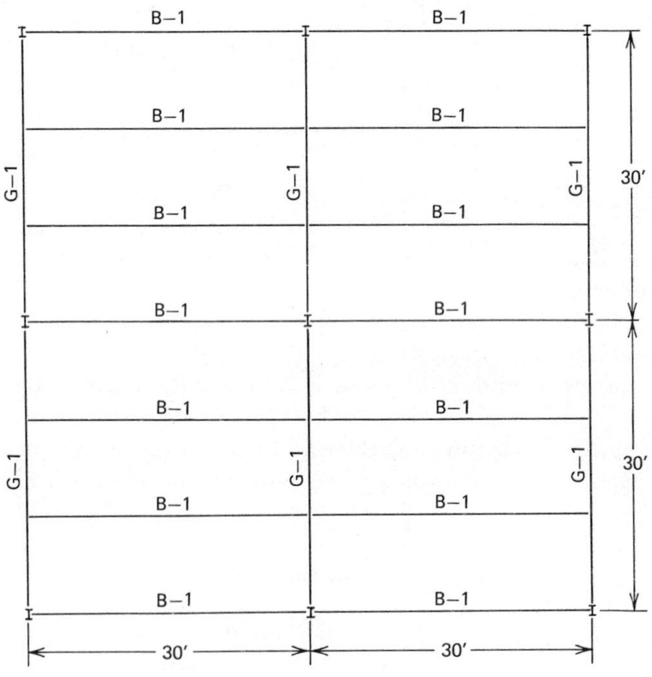

FIGURE 3.43 Beam and girder layout.

Determine the properties of the section:
Effective flange width (assume 10 in. flange)

$$b \leq \tfrac{1}{4}\,\text{span} = 90\text{ in.} \rightarrow b' = 40\text{ in.}$$

$$b' \leq \tfrac{1}{4}\,\text{clear distance} = 55\text{ in.}$$

$$b' \leq 8t = 48\text{ in. (for 6-in. slab)}$$

Use a 6-in.-thick slab, 10-in.-wide box girder, and a 90-in. effective flange width. Actually, this building plan is well suited to the use of composite metal deck floors as well as composite beams and girders. The metal-deck composite floor would easily span the 10 ft from beam to beam in the N-S direction, and the B-1 composite beams would frame to the box girder.

However, for purposes of the example, use the 6-in. concrete slab. Trial dimensions of the girder are shown in Figure 3.44.
Try $\frac{1}{2}$-in. top plate, 10 in. wide; two $\frac{5}{16}$-in. web plates, 24 in. deep; one $1\frac{1}{2}$-in. bottom plate, 10 in. wide.

$$M_T = M_D + M_L = 812.5\text{ kip-ft}$$

Use A-36 steel:

$$n = 9$$

$$f'_c = 3000\text{ psi}$$

$$f_c = 1350\text{ psi}$$

TRANSFORMED AREA OF SLAB $\quad \dfrac{6 \times 90}{9} = 60\text{ in.}^2$

$$\text{NA} = \frac{60(27) + (1/2 \times 10)(23.75) + (2)(5/16 \times 24)(12) + (1.5 \times 10)(0.75)}{60 + 5 + 15 + 15}$$

$$= \frac{1930}{95} = 20.32\text{ in.}$$

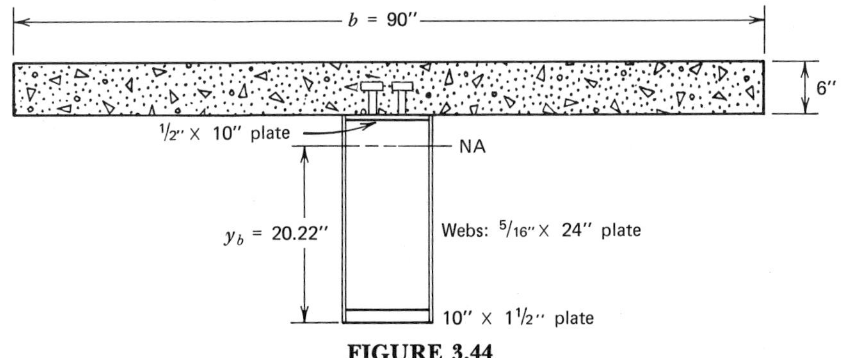

FIGURE 3.44

Moment of Inertia

Member	I_o	Ad^2
Slab	180	$60(6.68)^2 = 2677$
Top plate	—	$5(3.43)^2 = 59$
Web plates	720	$15(8.32)^2 = 1038$
Bottom plate	—	$15(19.57)^2 = 5745$
	900	9519

$$\text{Total } I = 900 + 9519 = 10{,}419 \text{ in.}^4$$

Preliminary Stress Check

$$f_s = \frac{(812.5)(12)(20.32)}{10{,}419} = 19 \text{ ksi}$$

This girder meets all the requirements of AISC Section 1.5.1.4.1, so that the allowable stress is 24 ksi. The girder as shown is about 20% overdesigned. A complete design would trim the depth somewhat for more efficient use of the material.

However, since this is only a demonstration problem, continue with the present dimensions

Properties of the Steel Section for Dead Load

$$\text{NA} = \frac{(1/2 \times 10)(23.75) + (2)(5/16 \times 24)(12) + (1.5 \times 10)(0.75)}{5 + 15 + 15} = 8.86 \text{ in.}$$

Moment of Inertia of Steel Section

Member	I_o	Ad^2
Top plate	—	$5(14.89)^2 = 1109$
Web plates	720	$15(3.14)^2 = 148$
Bottom plate	—	$15(8.11)^2 = 986$
	720	2243

Steel Stresses

BOTTOM PLATE $\dfrac{(276.6)(12)(8.86)}{2963} = 9.92 \text{ ksi}$ OK

TOP PLATE $\dfrac{(276.6)(12)(15.14)}{2963} = 16.96 \text{ ksi}$ OK

The compression flange of the girder is braced against lateral buckling by the B-1 beams framing in at 10-ft intervals.

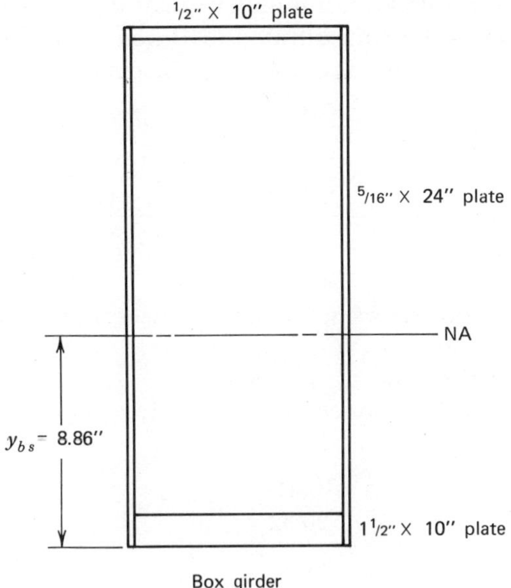

FIGURE 3.45 Box girder.

Check the concrete stress under total load:

$$f_c = \frac{(812.5)(12)(9.68)}{(9)(10{,}419)} = 1.0 \text{ ksi} < 1.35 \text{ ksi} \qquad \text{OK}$$

Shear Connection

The $\frac{1}{2}$-in. flange plate is thick enough to take a $\frac{7}{8}$-in. diameter stud. Use $\frac{7}{8} \times 3\frac{1}{2}$-in. studs in pairs.
Allowable load per stud = 15.6 kips.

$$V_h = \frac{0.85 f'_c A_c}{2} = \frac{(0.85)(3)(90 \times 6)}{2} = 688.5 \text{ kips}$$

$$V_h = \frac{A_s F_y}{2} = \frac{(35)(36)}{2} = 630 \text{ kips} \longleftarrow \text{use}$$

The number required = 630/15.6 = 40 studs on each side of the beam centerline.

This is the number of connectors that would be required if there were no concentrated loads applied to the beam. However, our girder has the B-1 beams framing in at 10-ft intervals and causing abrupt changes in the shear diagram of the beam, so AISC Formula 1.11-6 must be checked. This formula gives a number of shear connectors which must be spaced between the concentrated load point and the point of zero moment. In our case, the distance from the concentrated load point to the end of the girder span is 10 ft.

Formula 1.11.6

$$N_2 = \frac{N_1[(M\beta/M\text{max}) - 1]}{\beta - 1}$$

where

N_2 = number of connectors between the concentrated load and the zero moment point;
M = moment at the concentrated load;
N_1 = number of shear connectors previously computed;
$\beta = S_{tr}/S_s$.

$$S_{tr} = \frac{10{,}419}{20.32} = 512.75 \text{ in.}^3$$

$$S_s = \frac{2963}{8.86} = 334.42 \text{ in.}^3$$

$$\frac{S_{tr}}{S_s} = \frac{512.75}{334.42} = 1.53$$

M AT CONCENTRATED LOAD = 725 kip-ft
Mmax = 812.5 kip-ft

$$N_2 = \frac{40\{[725(1.53)/812.5] - 1\}}{1.53 - 1} = 28 \text{ connectors}$$

Figure 3.46 shows the number of studs required for the beam. Forty studs are required in each beam half (for a total of 80 studs). Of these 40 studs, 28 must be placed between the point of concentrated load and the end of the beam.

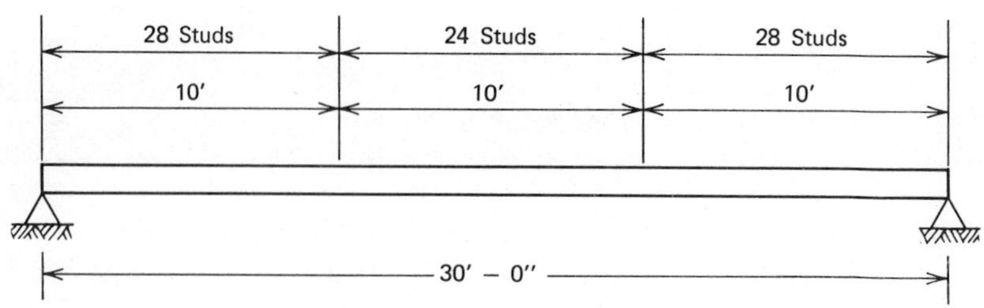

FIGURE 3.46 Stud spacing.

3.7 Compressive Steel within the Effective Slab Width

Shear connectors for composite beams are designed by the ultimate strength method. The ultimate capacity of the steel beam and the concrete slab is calculated. Then, using an appropriate factor to bring these two values into the working load range, the lower value is used to determine the number of connectors required.

The latest edition of the AISC specifications now recognizes that compression steel within the effective slab width raises the capacity of the slab. The formula for slab capacity then becomes

$$V_h = \frac{0.85 f'_c A_c}{2} + \frac{A'_s F_{yr}}{2}$$

where

A'_s = area of compression reinforcement;
F_{yr} = yield point of the reinforcement.

This addition becomes advantageous with shallow beams and especially beams with heavy coverplates. As an example, use a W 16 × 40 beam with a $1\frac{1}{2} \times 6$ coverplate. The effective flange width is 71 in. and the slab thickness is 4 in. Using No. 8 bars of Grade 60 steel at 6 in. *c/c*, 9.43 in.^{2} of steel is added to the slab.

Without the slab reinforcement, the horizontal shear capacity of the beam is the lower of

$$V_h = \frac{0.85 f'_c A_c}{2} = \frac{0.85(3)(4 \times 71)}{2} = 362 \text{ kips}$$

$$V_h = \frac{A_s F_y}{2} = \frac{(20.8)(36)}{2} = 374 \text{ kips}$$

Considering the slab reinforcement, the values become

$$V_h = \frac{0.85 f'_c A_c}{2} + \frac{A'_s F_{yr}}{2} = 362 + 282.8 = 644.8 \text{ kips}$$

$$V_h = \frac{A_s F_y}{2} = 374 \text{ kips, as above}$$

Also, changing the beam to 50-ksi steel would raise the capacity of the section by a great deal.

3.8 Partial Composite Action

In older editions of the specifications, partial composite action was referred to as incomplete composite action. The new term is preferred because it infers that the degree of composite action is under the control of the design engineer.

In actual practice with beam and slab construction, zero composite action is impossible because there is always some degree of bond and friction between the concrete slab and the steel beam. Similarly, 100% composite action is impossible because there is always some small degree of slip, no matter how rigidly the shear connection may be designed.

In very simplified form, Figure 3.47 shows the difference between full and partial composite action.

AISC actually recommends the use of partial composite action in order to obtain the maximum economy in design, because the partial composite action furnishes only the number of studs actually required to carry the load. The formulas for full composite action, which are based on ultimate strength, usually overestimate the number of studs and do not provide the most economical solution.

Partial composite action also makes possible a trade-off between beam depth, beam weight, and the number of studs required. Decreasing the number of studs naturally reduces the cost. However, if the solution is to use a deeper beam, the cost of exterior cladding material and pipe runs and ducts would nullify any advantage in a multistory building.

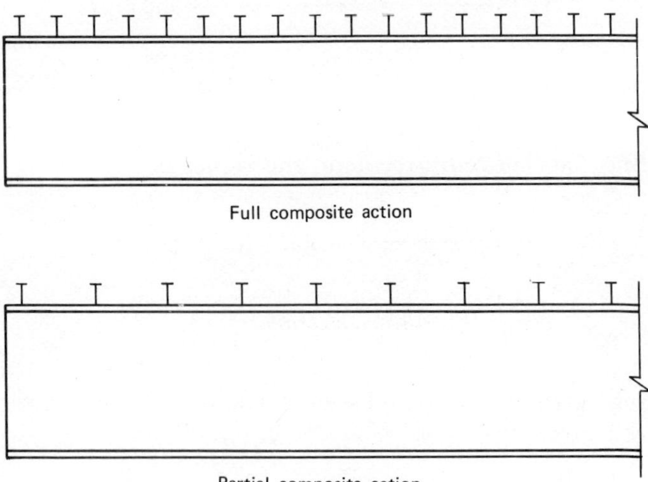

FIGURE 3.47 Partial composite action.

Decreasing the number of studs for partial composite action reduces the effective stiffness of the composite beam slightly so that live load deflections can be increased. However, the reduction from full to partial composite action generally does not have too much effect on deflections.

Recheck Example 3.2 for partial composite action:

$M_D = 56.3$ kip-ft REQUIRED $S_{tr} = 117.8$ in.^{3}

$M_L = 179.2$ kip-ft REQUIRED $S_s = 28.2$ in.^{3}

Section selected:

W 18 × 50 with 4 in. slab

$b = 71.5$ in.
$y_b = 16.52$ in.
$S_{tr} = 125$ in.^{3}
$V_h = 265$ kips
$S_t = 376$ in.^{3}
$S_s = 89.1$ in.^{3}
$I_{tr} = 2060$ in.^{4}

24 studs on each side of the beam centerline for a total of 48 studs

ALLOWABLE LIVE LOAD DEFLECTION $\dfrac{L}{360} = 1.06$ in.

ACTUAL LIVE LOAD DEFLECTION 0.55 in.

For partial composite action:

$$S_{eff} = S_s + \frac{V_h'}{V_h}(S_{tr} - S_s)$$

where V_h' equals the stud value times the number of studs actually used on each side of the beam centerline. S_{eff} is the value of the transformed section modulus required to carry the load, not the section modulus supplied.

Solving the above equation for V_h':

$$V_h' = V_h\left[\frac{S_{eff} - S_s}{S_{tr} - S_s}\right]$$

$$V_h' = 265\left[\frac{117.8 - 89.1}{125 - 89.1}\right] = 212 \text{ kips}$$

NUMBER OF STUDS $\dfrac{212}{11.5} = 18.4$

Use 19 studs on each side of the beam centerline for a total of 38 studs

SAVINGS $48 - 38 = 10$ studs per beam

This beam provides 80% composite action. The AISC specifications set a lower limit of 25%, but it is seldom practical to provide less than 50% partial composite action.

Live Load Deflection Check

Supplement No. 3 to the AISC specifications gives a formula for a reduced effective moment of inertia for partial composite action.

$$I_{eff} = I_s + \sqrt{\frac{V'_h}{V_h}}(I_{tr} - I_s)$$

$$I_{eff} = 802 + \sqrt{\frac{212}{265}}(2060 - 802)$$

$$I_{eff} = 1927 \text{ in.}^4$$

DEFLECTION $\quad \dfrac{2060}{1927}(0.55) = 0.59 \text{ in.} < 1.06 \text{ in.}$

Now try a deeper beam with partial composite action. Although partial composite action reduces stiffness, the deeper beam will increase stiffness so that deflections will probably not be critical.

From AISC composite beam tables, select the following:

W 21 × 49 (keeping beam weight virtually the same)

$$S_{tr} = 134 \text{ in.}^3$$
$$S_t = 429 \text{ in.}^3$$
$$S_s = 93.3 \text{ in.}^3$$
$$I_{tr} = 2530 \text{ in.}^4$$

Because of a different steel area, use a new V_h:

$$V_h = \frac{A_s F_y}{2} = \frac{14.4(36)}{2} = 2.59 \text{ kips}$$

$$V'_h = V_h \left[\frac{S_{eff} - S_s}{S_{tr} - S_s}\right]$$

$$= 259 \left[\frac{117.8 - 93.3}{134 - 93.3}\right] = 156 \text{ kips}$$

NUMBER OF STUDS, N_s $\quad \dfrac{156}{11.5} = 13.6$

Use 14 studs on each side of the beam centerline for a total of 28 studs

SAVINGS $\quad$ = another 10 studs per beam

Live Load Deflection Check

$$I_{\text{eff}} = I_s + \sqrt{\frac{V'_h}{V_h}}(I_{tr} - I_s)$$

$$= 971 + \sqrt{\frac{156}{259}}(2530 - 971) = 2180 \text{ in.}^4$$

ACTUAL DEFLECTION $= \dfrac{2060}{2180}(0.55) = 0.52$ in.

Because of the deeper beam, this is actually less than the deflection of the W 18 × 50 beam.

3.8.1 Cost Comparisons—Partial Composite Action

The four previous beam systems are now compared. The composite systems are compared to a W 18 × 64 noncomposite beam that carries the same load.

This is an office building in the 7 to 15 story range. The comparison is for one bay of a multibay building. The bay size is 32 × 32 ft, with four beams per bay. Costs used are from 1974 figures.

Beam	Steel Weight	Steel Cost	Cost of Studs	Total	Savings
W 18 × 64; Noncomposite	4.1 tons	$2337	—	$2337	—
W 18 × 50; Full composite	3.2 tons	$1824	$115	$1939	17%
W 18 × 50; Partial composite	3.2 tons	$1824	$ 91	$1915	18%
W 21 × 49; Partial composite	3.1 tons	$1788	$ 67	$1855	21%

The reduction in steel tonnage of the beams means smaller loads to the columns and foundations. In this particular building, the live-load to dead-load ratio is fairly high, so that column sizes would be affected only very slightly.

3.9 The Use of Lightweight Concrete

Concrete made from lightweight aggregates can be used satisfactorily in composite beams. There is no reason why lightweight concrete cannot be combined with materials other than steel. However, only the steel specifications give a special requirement for the shear connection when lightweight concrete is used.

The lightweight concretes range from 90 to 120 pcf in unit weight. They utilize man-made aggregates such as expanded shale or expanded slate which conform to ASTM designation C330.

For computations of stress, the modular ratio for normal weight concrete can be used for both normal and lightweight concrete. However, for deflection computations, the actual value of n for the lightweight concrete must be used. With lightweight concrete, the effects of creep and shrinkage increase substantially. Long-term deflections that use a value of $2n$ for calculation naturally show a marked increase. In the case of Example 4.2, the long-term deflections using lightweight concrete are 50% higher than the short-term deflections using stone concrete.

Shear connector strengths are reduced when lightweight concretes are used. Table 1.11.4A from Supplement 3 to the AISC specifications, which is reproduced here as Table 3.2, gives coefficients that are to be multiplied by the standard connector strengths to give the appropriate value for use with lightweight concrete. Research has shown that the modulus of elasticity and the ultimate compressive strength of the concrete are the controlling factors in determining shear connector strength.

Coefficient values for concrete strengths between 4.0 and 5.0 ksi can readily be interpolated from the table.

This reduced value of shear connector strength does not mean weaker composite beams when lightweight concrete is used. It simply means that a few more studs are required to obtain full composite action. The ultimate carrying capacity of the beam depends on concrete strength, not concrete weight. Weight of the concrete makes no difference in the ultimate carrying capacity of the beam, providing the shear connection is properly designed.

Composite beams with lightweight concrete can also be designed for partial composite action, but the use of less than 50% composite action is not recommended.

Lightweight concrete lends itself well to unshored construction be-

Table 3.2

Air Dry Unit Weight, pcf	90	95	100	105	110	115	120
Coefficient, $f'_c \leq 4.0$ ksi	0.73	0.76	0.78	0.81	0.83	0.86	0.88
Coefficient, $f'_c \geq 5.0$ ksi	0.82	0.85	0.87	0.91	0.93	0.96	0.99

cause the lighter concrete can reduce the slab dead load by as much as 35%. For dead load deflection calculations, less dead load means smaller deflections, but the higher n value causes larger deflections. These two effects approximately cancel out each other so that the net effect is very little change in dead load deflections when lightweight concrete is used. These calculations are shown in detail in Chapter 8.

Check of Example 3.2 Using Lightweight Concrete

Use concrete with a unit weight of 110 pcf and an ultimate strength of 3.0 ksi.

$$\text{MODULUS } E_c = w^{1.5}(33\sqrt{f'_c})$$

$$= (110)^{1.5}(33\sqrt{3000})$$

$$= 2{,}085{,}000 \text{ psi}$$

$$n = \frac{E_s}{E_c} = \frac{29 \times 10^6}{2.085 \times 10^6} = 14$$

$$\text{DL} = 8 \times \frac{4}{12} \times 110 = 293$$

$$\text{ASSUMED BEAM} \qquad = \underline{45}$$

$$338 \text{ p.p.f.}$$

$$M_D = \frac{0.338(32)^2}{8} = 43 \text{ kip-ft}$$

From a previous example,

$$M_L = 179.2 \text{ kip-ft.}$$

$$M_T = M_D + M_L = 43 + 179.2 = 222.5$$

$$\text{REQUIRED } S_{tr} = \frac{22.5(12)}{24} = 111.25 \text{ in.}^3$$

From composite beam tables, select W 18 × 45:

$$S_{tr} = 112 \text{ in.}^3 \qquad S_s = 79.0 \text{ in.}^3 \qquad y_b = 16.65$$

$$S_t = 358 \text{ in.}^3 \qquad I_{tr} = 1860 \text{ in.}^4 \qquad b = 71.5$$

$$\text{CONCRETE STRESS} \quad \frac{222.5(12)}{358(9)} = 0.83 \text{ ksi} < 1.35 \text{ ksi}$$

Shear Connection

$$V_h = \frac{0.85(3)(4 \times 71.5)}{2} = 364.6 \text{ kips}$$

$$V_h = \frac{13.2(36)}{2} = 237.6 \text{ kips} \longleftarrow \text{controls}$$

STUD CONNECTOR VALUE $\quad 0.83(11.5) = 9.5$ kips

NUMBER OF STUDS $\quad \frac{237.6}{9.5} = 25$ studs on each side of the beam centerline.

The tabulation below shows the comparison between the beams with normal weight and lightweight concrete.

	Beam Size	Studs Required	Δ_{DL}	Δ_{LL}
Normal Weight Concrete	W 18 × 50	46	0.45 in.	0.55 in.
Lightweight Concrete	W 18 × 45	50	0.39 in.	0.67 in.

3.10 AISC Shoring Formula

Formula 1.11-2 in the AISC specifications states that for construction without shoring, the tensile section modulus of the beam shall not exceed

$$S_{tr} = \left[1.35 + 0.35\frac{M_L}{M_D}\right] S_s \tag{3.1}$$

Unshored steel-concrete composite members used to be designed by a two-step process. Stresses in the steel due to dead load were computed from M_D/S_s. These were added to stresses in the composite section, caused by live load, and were computed from M_L/S_{tr}. The total stress in the bottom flange was the sum of

$$f_s = \frac{M_D}{S_S} + \frac{M_L}{S_{tr}} \tag{3.2}$$

Research has shown that there is no difference in the ultimate carrying capacity of shored and unshored members, so that unshored members can be designed according to

$$f_s = \frac{M_D + M_L}{S_{tr}} \tag{3.3}$$

This research recognizes that there is a high factor of safety of the composite member against ultimate failures. Thus it is reasonable to allow an overstress of 35% in Equation 3.3, so that this equation becomes

$$f_s = 1.35 \frac{M_D + M_L}{S_{tr}}$$

By equating this value to Equation 3.2, the AISC equation is easily derived. In short then, the AISC formula allows a reasonable overstress in the beam, but it still limits the ratio of S_{tr}/S_s for shoring requirements.

Using a live-load to dead-load ratio of 3 : 1, the AISC formula reduces to

$$S_{tr} = 1.35 + 0.35 \left(\frac{3}{1}\right) S_s$$

from which

$$\frac{S_{tr}}{S_s} = 2.40$$

A quick check of a dozen or so sections which might be typical composite beams, and with live-load to dead-load ratios varying from 1 : 1 to 3 : 1, shows that few sections fail to meet this requirement. The check shows us that this code requirement, while it limits the ratio of S_{tr} to S_s, is not too restrictive for practical design purposes.

However, this formula relates to the bottom fibers of the beam which are in tension. The formula does not relieve the designer and contractor of providing adequate support for the compression flange of the beam during construction.

References

3.1 Lothers, J. E., *Advanced Design in Structural Steel*, Prentice-Hall, New York, 1960.

3.2 *Manual of Steel Construction*, 7th ed., American Institute of Steel Construction, New York, 1973.

3.3 Sherman, J., "Continuous Composite Steel and Concrete Beams," *ASCE Transactions*, No. 119, 1954.

3.4 Iwamoto, K., "On the Continuous Composite Girder," Highway Research Board Bulletin No. 339, National Academy of Science, Washington, D.C., 1962.

3.5 "Composite Construction for Structural Steel and Concrete, Beams for Bridges," British Standard Code of Practice, British Standards Institution, London, 1967.

3.6 McCormac, Jack C., *Structural Steel Design*, Intext Publishers, New York, 1971.

3.7 Slutter, R. G., and Driscoll, G. C., "Flexural Strength of Steel-Concrete Composite Beams," *J. Str. Div. ASCE*, April 1965.

3.8 "Progress Report of the Joint ASCE–ACI Committee on Composite Construction," *J. Str. Div. ASCE,* December 1960.

3.9 Slutter, R. G., "Composite Steel Concrete Members," *Structural Steel Design,* L. S. Beedle, Ed., Ronald Press, New York, 1964.

3.10 Knowles, P. R., *Composite Construction in Steel and Concrete,* Halsted Press, 1973.

3.11 Viest, I. M., Fountain, R. S., and Singleton, R. C., *Composite Construction in Steel and Concrete,* McGraw-Hill, New York, 1958.

3.12 Sherman, J., "Continuous Composite Steel and Concrete Beams," *ASCE Transactions,* No. 119, 1954.

3.13 Newmark, N. M., Siess, C. P., and Viest, I. M., "Studies of Slab and Beam Highway Bridges, Part III, Small Scale Tests of Shear Connectors and Composite T-Beams," University of Illinois, Eng. Exp. Station Bulletin No. 396, 1952.

3.14 *Alpha Composite Construction Engineering Handbook,* Poiete Manufacturing Company, North Arlington, N.J., 1949.

3.15 McGarraugh, J. B., and Baldwin, J. B. W., "Lightweight Concrete on Steel Composite Beams," *AISC Engineering Journal,* Vol. 8, No. 3, 1971.

4

Concrete-Concrete Composite Beams

The concrete-concrete composite beam, like other composite beams, consists of a beam and slab tied together to act as a unit. The shear connection between the beam and slab should tie the components together well enough so that they act as a monolithic T beam.

The beam can be either a cast-in-place or precast unit. Since it is easier to obtain better quality control of the concrete in the casting yard, it is not uncommon to use different strengths of concrete for the beam and slab if a precast unit is to be used.

In general, cost leads to the use of the precast beam unit. The high cost of construction is largely due to the cost of skilled on-site labor. The precasting yard can use a great deal of local semiskilled labor and reuse forms which can lower costs. The precast beam also means shorter construction time, but it is subject to the hazards of transporting and delivery schedules.

The precast members can also be built to closer tolerances. Table 4.1 which was compiled by the British Building Research Station, shows a comparison of precasting and on-site construction.

The economics of using a high-strength precast beam is open to question. After the slab is cast, the neutral axis of the composite beam is usually close to the junction of the beam and slab, so that little, if any, of the higher strength (and higher price) concrete actually helps the composite member. One factor does lead to the use of high-strength beam concrete. The ACI code states that the individual elements shall be investigated for all critical stages of loading, so that the beam member must be checked for permanent dead load and other construction loads.

Design of the composite concrete-concrete member should be in accordance with Chapter 17 of ACI Code 318-71. Other sections of the

Table 4.1

Type of Production	Site Casting	Normal Factory Casting	Well-Controlled Factory Casting
Tolerance in mold fabrication	±0.12 in.	±0.06 in.	±0.02 in.
Tolerance in slabs cast	±0.32 in.	±0.20 in.	±0.08 in.
Tolerance in erection	±0.24 in.	±0.20 in.	±0.16 in.
TOTAL TOLERANCE	±0.40 in.	±0.28 in.	±0.18 in.

code which affect composite members are noted in Chapter 9 which includes control of deflections and the sections on shear-friction (Section 11.15), T beams (Section 8.7), and deep beams (Sections 10.7 and 11.9).

Composite members may consist of combinations of high-strength and low-strength concrete, precast and cast-in-place units, and normal weight and lightweight concrete. These are some of the combinations available in T beams, so that the designer has a great deal of flexibility in the design, proportioning, and construction of composite members. The usual case is the precast beam with a cast-in-place slab to form a T section.

Prestressing can be used effectively with composite beams. Either the precast stem can be prestressed or the slab can be cast and composite action established, and then the entire unit can be prestressed.

4.1 The Shear Connection

The ACI code states that if the following four criteria are met, horizontal shear need not be calculated.

1. Contact surfaces between the beam and slab are clean and intentionally roughened.
2. Minimum ties are provided.
3. Web members are designed to resist the entire vertical shear.
4. All stirrups are fully anchored into all intersecting components.

Full transfer of the shear forces should be provided at the interface.

Note that there is no provision for partial composite action, as in the AISC code. If the four requirements above are met, then full transfer of the horizontal shear can be assumed.

If the horizontal shear is to be calculated, the computation is quite simple. The horizontal shear is found from

$$v_{dh} = \frac{V_u}{\phi b_v d} \tag{4.1}$$

where v_{dh} = design horizontal shear at any cross section in psi;
V_u = total applied design shear force;
ϕ = capacity reduction factor (0.85 for shear);
b_v = width of the interface cross section;
d = distance from the compression face to the tension reinforcement. The dimension d is for the entire composite section.

Horizontal shear is transferred by direct bond, shear friction, and reinforcement across the interface. At working loads, the direct bond is usually intact. Shear friction becomes active when the bond is broken, as by cyclic loading, or at high load levels.

The ACI code allows a permissible value of 80 psi in horizontal shear when the contact surfaces of the components are clean and "intentionally roughened," but ties are not provided.

When ties are provided, but the surfaces are not roughened, the same value of 80 psi is allowed.

When both ties and intentional roughening are provided, the code allows a design value of 350 psi.

If the design horizontal shear exceeds 350 psi, the design for horizontal shear must be done in accordance with the section of the ACI code on shear friction (Section 11.15).

Intentional roughness may be assumed only when the interface is roughened with a full amplitude of $\frac{1}{4}$ in. Methods of obtaining this intentional roughness are spelled out in Chapter 7, "The Shear Connectors."

Practically speaking, a complete design for the beam stem will include vertical ties or stirrups for web reinforcement. All that remains is to detail the ties so that they extend far enough up into the slab to be effective in transmitting the horizontal shear. The shear connector check then reduces to a simple check of whether the intentional roughening is required to supplement the shear reinforcement.

The vertical ties can have any of several configurations, but the closed loop is usually considered the most effective because it provides very good uplift resistance. This assumes that there is enough room to provide the full embedment length of the bar.

4.2 Effective Flange Width

The object in designing the composite member is to have a member at least as strong as a monolithic T beam. Consequently, the restrictions on the effective flange width of T beams also apply to composite beams.

The ACI code specifies temperature and shrinkage reinforcement in slabs where the principal reinforcement extends in one direction only. In a composite beam system, the slab is designed as a one-way slab spanning transversely between the beams. The temperature and shrinkage steel required by the code, in addition to serving its primary function, is quite effective in maintaining the structural integrity of the slab overhang portion of the T beam. The code limits the spacing of temperature steel to 5 times the slab thickness, or 18 in.

It should be obvious that if the beams are spaced quite far apart, as in Figure 4.1, the entire slab will not be effective as a compression flange for the beam. This same figure should also make it apparent that the slab thickness and stem width also affect the effective flange width. The code spells out the flange width requirements as follows:

"For symmetrical Tee beams, the overhang on either side of the web shall not exceed 8 times the slab thickness, nor one-half the clear distance to the next beam. The effective flange width shall not exceed one-fourth the span length of the beam."

For design then, the effective flange width shall be the least of the following three values:

1. Width of stem plus 16 times the slab thickness.
2. Distance center to center of beams.
3. One-fourth of the span length.

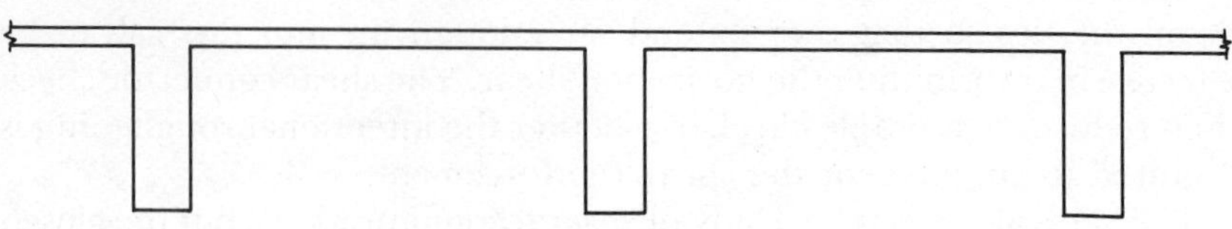

FIGURE 4.1 Poor proportions.

4.3 Example Problem-Strength Method

The 1971 ACI code generally recommends the use of the ultimate strength design method, but it recognizes and accepts an alternate method, which is the older working stress method of design.

In the design of the shear connection, the code specifies the use of $v = V/\phi bd$ for determining shearing stress by the ultimate design method. In the working stress method, the allowable stresses are reduced to 55% of the code values, and the ϕ and U factors are not used. Wang and Salmon (4.5) effectively show that there is no real advantage in using an alternate method for shear because the combination of reduced stress, U and ϕ, results in little, if any, change in the shear requirements.

Example 4.1

SPAN	24 ft	f'_c slab = 3,000 psi	$f_c = 1{,}350$ psi
		f'_c beam = 4,000 psi	$f_c = 1{,}800$ psi
		f_y steel = 60,000 psi	$f_s = 24{,}000$ psi

SLAB THICKNESS 4 in.

Use a precast beam with $f'_c = 4000$ psi and a cast-in-place slab with a strength of 3000 psi.

Effective Width

c/c beams	$= 6 \times 12$	$= 72$ in.
1/4 span	$= 1/4(24 \times 12)$	$= 72$ in.
$16t + b_f$	$= 16t + 8$ (assumed)	$= 72$ in.

The 6-ft beam spacing and 4-in. slab thickness seem to have been arbitrarily assumed. Actually, a great deal of construction planning is necessary in order to make these decisions. Naturally, a smaller spacing for the beams means a shorter transverse slab span and a thinner slab. Also, a concrete beam is usually in a good economic range if the beam is roughly twice as deep as it is wide. Balanced against this fact is the total depth of the floor system. Just 2 in. saved in the depth of the floor system can mean a lot of money.

Forming of the slab should also be considered. The bottom form for the slab is usually cut from 4 × 8-ft plywood sheets. Careful planning of beam spacings can mean savings in the cost of formwork.

In an actual job, two or three possible combinations of beam sizes and spacings should be roughed out to determine the best cost factor.

Loads on Noncomposite Precast Beam

DL, SLAB	$\frac{4}{12} \times 6 \times 0.15 = 0.30$ kpf
BEAM STEM $(8 \times 12$ est.)	$\frac{8}{12} \times \frac{12}{12} \times 0.15 = \underline{0.10 \text{ kpf}}$
PERMANENT DEAD LOAD	$= 0.40$ kpf
TEMPORARY CONSTRUCTION LOAD	$= \frac{50 \text{ psf}}{1000} \times 6 = 0.3$ kpf
LIVE LOAD	$\frac{150 \text{ psf}}{1000} \times 6 \quad = 0.9$ kpf

The 50 psf of temporary construction load is not an estimated figure. Many codes specify a superimposed live load for falsework and other temporary construction conditions. As just one example, Section 403.3.5 of the Kentucky Department of Highways Standard Specifications contains a provision of 50 psf for these conditions.

Use the ultimate strength design method. Consider the construction load as live load on the precast beam.

$$M_D = \frac{0.4 \times (24)^2}{8} = 28.8 \text{ kip-ft}$$

$$M_L = \frac{0.3 \times (24)^2}{8} = 21.6 \text{ kip-ft}$$

$$U = 1.4(28.8) + 1.7(21.6) = 40.32 + 36.72 = 77 \text{ kip-ft}$$

$$\overline{M_u} = \frac{77}{0.9} = 85.6 \text{ kip-ft}$$

The percentage of steel in the beam should be less than $0.75p_b$ and greater than $200/F_y$. Ordinarily, it is best to stick to these limits. However, when building the precast section, enough steel must be placed to take care of the live load requirements of the composite section, so that the precast unit may often be over reinforced.

For our beam use the maximum steel percentage.

$$p = 0.75p_b$$

$$p_b = \frac{(0.85)^2 f'_c}{F_y} \left(\frac{87{,}000}{87{,}000 + F_y} \right)$$

$$p_b = \frac{(0.85)^2(4)}{60} \left(\frac{87{,}000}{147{,}000} \right) = 0.0285$$

$$0.75p_b = 0.75(0.0285) = 0.021 \longleftarrow \text{use}$$

The loads on our beam are not too heavy, so the steel will probably be placed in a single row. Minimum stem width then becomes a practical consideration. At the beginning of this example, an 8-in. stem width was assumed. Within this stem width, three bars of a practical size cannot be placed, so that the most probable steel selection is two bars in the No. 9, 10, or 11 sizes.

From any reinforced concrete text, we find

REQUIRED $$Mu = pF_y bd^2 \left(1 - 0.59p \frac{F_y}{f_c}\right)$$

from which

$$M_u = 0.923 bd^2 \qquad \text{for our case}$$

$$bd^2 = \frac{M_u}{0.923} = \frac{77 \times 12}{0.923} = 1001 \text{ in.}^3$$

Using an 8-in. stem width,

$$d = 11.2 \text{ in.}$$

Assuming No. 10 bars, the total stem depth is

$$11.2 + \frac{1.27}{2} + 0.375 + 1.5 = 13.71 \text{ in.}$$

Use full inch dimensions for the stem, so that the actual size is 8×14 and the actual "d" is

$$14 - 1.5 - 0.375 - \frac{1.27}{2} = 11.5 \text{ in.}$$

Two No. 10 bars furnish 2.54 in.^{2} of steel.

Static Check of the Precast Section

$$T = A_s F_y = 2.54(60) = 152.4 \text{ kips}$$

Depth of stress block:

$$a = \frac{T}{(0.85)(f'_c)(b)} = \frac{152.4}{(0.85)(4)(8)} = 5.60 \text{ in.}$$

$$\overline{Mu} = T\left(d - \frac{a}{2}\right) = \frac{152.4}{12}\left(11.49 - \frac{5.60}{2}\right) = 110.4 \text{ kip-ft} > 85.6 \text{ kip-ft}$$

Check T Section

PERMANENT DL, SLAB $= 0.30$ kpf

BEAM STEM $\left(\frac{8}{12}\right)\left(\frac{14}{12}\right)(0.15) = 0.12$ kpf

$$M_D = \frac{0.42 \times (24)^2}{8} = 30.24 \text{ kip-ft}$$

$$M_L = \frac{0.9 \times (24)^2}{8} = 64.8 \text{ kip-ft}$$

$$\begin{aligned} U &= 1.4D + 1.7L \\ &= 1.4(30.24) + 1.7(64.8) \\ &= 42.34 + 110.16 = 152.5 \text{ kip-ft} \end{aligned}$$

$$\overline{M_u} = \frac{U}{\phi} = \frac{152.5}{0.9} = 169.4 \text{ kip-ft}$$

Check to see whether the neutral axis is in the slab or the stem. Assume that the neutral axis is at the junction of the slab and stem. Then compute the tensile and compressive forces.

$$\begin{aligned} C &= 0.85(f'_c)(B_1)(b)(t) \\ &= 0.85(3)(0.85)(72)(4) = 624.2 \text{ kips} \\ T &= A_s F_y = 2.54(60) = 152.4 \text{ kips} \end{aligned}$$

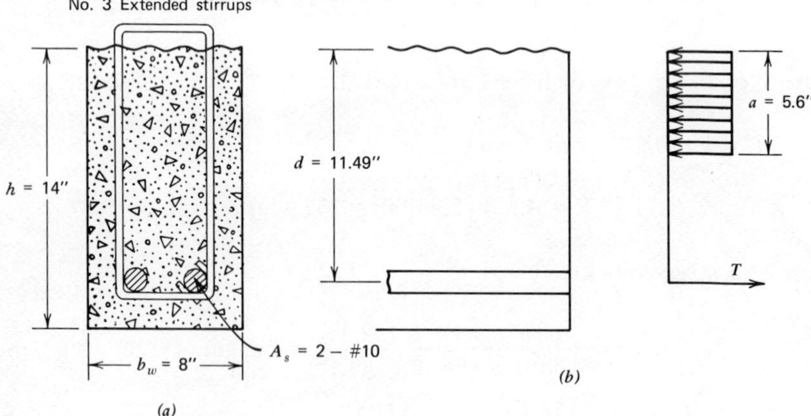

FIGURE 4.2 Cross section of precast beam.

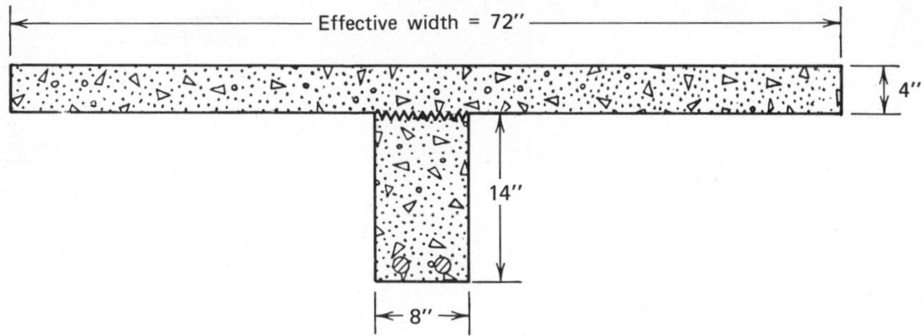

FIGURE 4.3 Composite T beam.

The compressive factor (the slab) is larger, so obviously the entire depth of the slab cannot be in compression. Now figure the actual depth of the compressive stress block.

$$a = \frac{T}{(0.85)(f'_c)(b)} = \frac{152.4}{(0.85)(3)(72)} = 0.83 \text{ in.}$$

MOMENT ARM $= d - \frac{a}{2} = 15.49 - \frac{0.83}{2} = 15.08 \text{ in.}$

$$\overline{Mu} = T \times \text{moment arm} = \frac{152.4}{12}(15.08) = 191.5 \text{ kip-ft} > 169.4 \text{ kip-ft}$$

Check the shear connection:

$$V_D = 0.42 \text{ kpf} \times \frac{24}{2} = 5.04 \text{ kips}$$

$$V_L = 0.9 \text{ kpf} \times \frac{24}{2} = 10.8 \text{ kips}$$

$$V_u = 1.4D + 1.7L$$

$$= 1.4(5.04) + 1.7(10.8)$$

$$= 7.06 + 18.36 = 25.42 \text{ kips}$$

$$= \frac{V_u}{\phi b_w d} = \frac{25{,}420}{(0.85)(8)(15.49)} = 240 \text{ psi}$$

This is greater than 80 psi, so that both extended stirrups and intentional roughening are required.

REQUIRED STIRRUPS $A_v = \frac{50 b_w S}{F_y}$

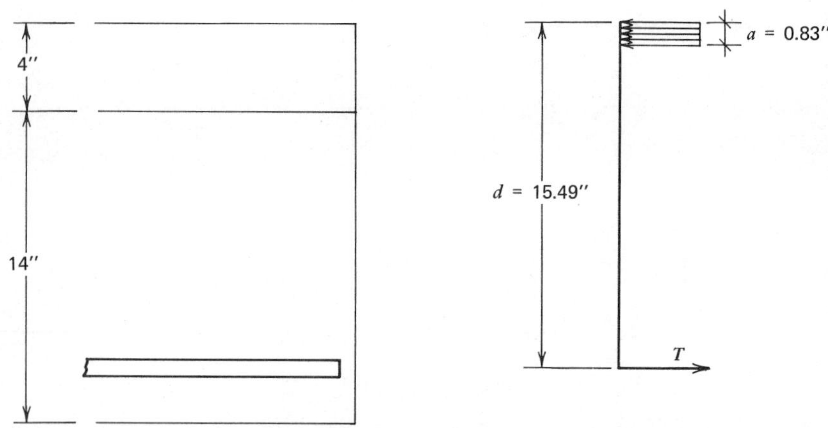

FIGURE 4.4 Forces of ultimate load.

Using No. 3 bars, two bar areas cross the slab-stem interface for each stirrup.

$$A_v = 2(0.11) = 0.22 \text{ in.}^2$$

Solve for spacing:

$$S = \frac{A_v F_y}{50 b_w} = \frac{(0.22)(60{,}000)}{(50)(8)} = 33 \text{ in.}$$

By ACI Code 11.1.4, stirrup spacing is limited to $d/2$. Unless a closer spacing of the web reinforcement has been previously designed for the precast member, use $15.08/2 = 7\frac{1}{2}$ in.

4.4 Example Problem—Alternate Method

The same beam is to be redesigned using the working stress method.

Although the value of effective flange width does change somewhat with load, the ACI code recommends the use of the same effective flange width for both design methods.

As before, effective flange width = 72 in.

f'_c slab = 3,000 psi	f_c slab = 1,350 psi
f'_c beam = 4,000 psi	f_c beam = 1,800 psi
F_y steel = 60,000 psi	f_s steel = 24,000 psi

Loads on Noncomposite Precast Beam

The alternate method will probably produce a larger beam section, so a larger estimated beam weight will be used.

SLAB $\left(\frac{4}{12}\right)\times 6\times 0.15 = 0.30 \text{ kpf}$

BEAM STEM (12 × 16 est.) $\left(\frac{12}{12}\right)\times\left(\frac{16}{12}\right)\times 0.15 = \underline{0.20 \text{ kpf}}$

0.50 kpf

TEMPORARY CONSTRUCTION LOAD $\frac{50 \text{ psf}}{1000}\times 6 = 0.3 \text{ kpf}$

Consider the construction load as live load on the precast beam.

$$M_D = \frac{0.50\times(24)^2}{8} = 36 \text{ kip-ft}$$

$$M_L = \frac{0.3\times(24)^2}{8} = 21.6 \text{ kip-ft}$$

Live Load On Composite Beam

As before $M_L = 64.8$ kip-ft

Design Constants

BEAM $f'_c = 4000$ psi $n = 8$
SLAB $f'_c = 3000$ psi $n = 9$

Precast Beam Design

$$k = \frac{f_c}{(f_s/n)+f_c} = \frac{1{,}800}{(24{,}000/8)+1{,}800} = 0.375$$

$$j = 1-\frac{k}{3} = 1-\frac{0.375}{3} = 0.875$$

The resisting moment of the section equals either T or C times the moment arm of the couple.

MOMENT ARM $jd = d - \frac{kd}{3}$

$$M_r = \tfrac{1}{2} f_c jkbd^2$$

Find the size of the member using the resisting capacity of the concrete compression area. The amount of steel will be determined by the requirements of the composite section.

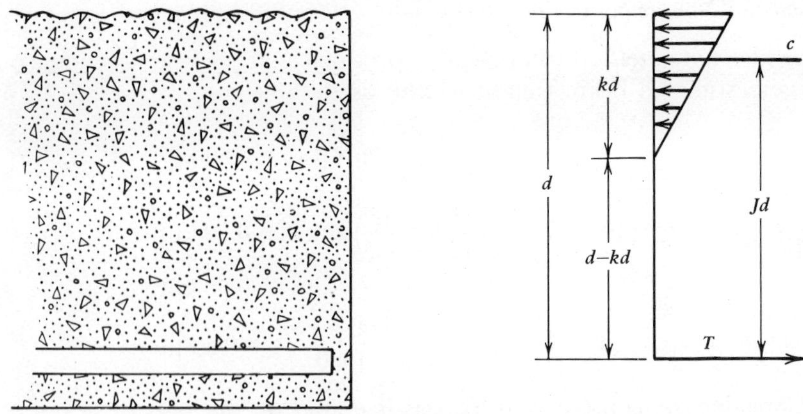

FIGURE 4.5 Beam stresses—alternate method.

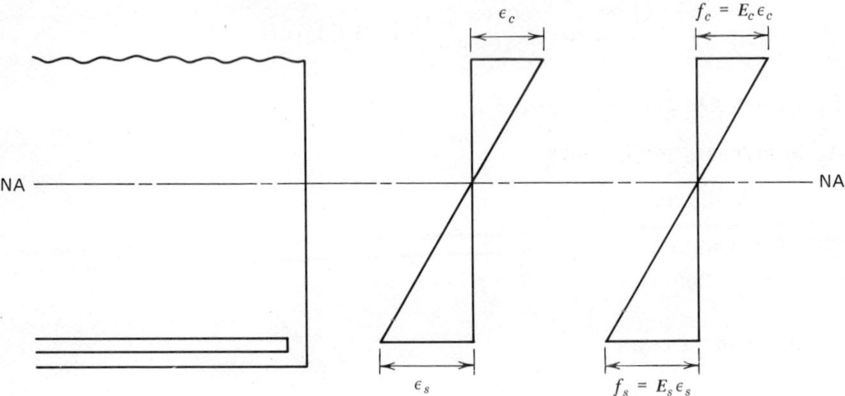

FIGURE 4.6 Stress and strain distributions.

TOTAL MOMENT ON THE PRECAST BEAM $36 + 21.6 = 57.6$ kip-ft

$$bd^2 = \frac{M}{1/2f_cjk} = \frac{57.6(12{,}000)}{1/2(1{,}800)(0.875)(0.375)} = 2{,}340 \text{ in.}^3$$

Construction practice is to keep beam stem widths in multiples of 2 or 3 in. and overall depths to even inches, with

$$12 \text{ in. width} - \text{required } d = 13.96 \text{ in.}$$
$$10 \text{ in. width} - \text{required } d = 15.29 \text{ in.}$$

Try the 10-in. stem width. This will accommodate three bars up to the

No. 9 size in a single row. Total height of the stem will be approximately $2\frac{1}{2}$ in. larger than the required d. Try a beam stem size of 10×18 in.

Assuming No. 9 bars and No. 3 stirrups,

ACTUAL $$d = 18 - 1.5 - 0.375 - \frac{1.128}{2} = 15.56 \text{ in.}$$

The actual beam weight is close enough to the assumed weight, so that the dead load need not be revised.

Reinforcing Steel

$$\begin{aligned} A_s &= \frac{M_D}{f_s jd \text{ (precast)}} + \frac{M_L}{f_s jd \text{ (composite)}} \\ &= \frac{36(12{,}000)}{24{,}000(0.875)(15.56)} + \frac{64.8(12{,}000)}{24{,}000(0.875)(19.56)} \\ &= 1.32 + 1.89 = 3.21 \text{ in.}^2 \end{aligned}$$

Three No. 9 bars only furnish 3.00 in.^{2} of steel, so the beam is unsatisfactory. A wider beam would mean using a 12-in. width. This would, however, permit the placing of three No. 10 bars in one row. Beam depth can be increased in 1-in. increments, and an increase in beam depth increases the moment arm of the resisting couple. Either choice is open. Both should be checked for overall cost in the building.

In our case, try to increase the stem size to 10×19 in. Then $d = 16.56$-in. Stick with three No. 9 bars, $A_s = 3.0$ in.^{2}

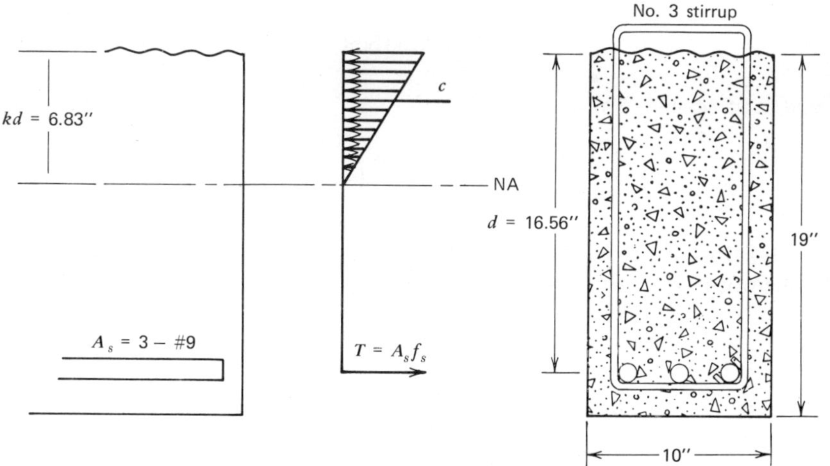

FIGURE 4.7 Precast beam—alternate method.

Check the precast noncomposite section first.

$$d = 16.56 \qquad n = 8$$

Locate the neutral axis. Summing moments about the neutral axis:

$$b(kd)\left(\frac{kd}{2}\right) = nA_s(d - kd)$$

$$10(kd)\left(\frac{kd}{2}\right) = 8(3.0)(16.56\ kd)$$

from which $kd = 6.83$ in.:

$$k = \frac{kd}{d} = \frac{6.83}{16.56} = 0.41 \qquad j = 1 - \frac{k}{3} = 0.86$$

Note that these values of j and k are different from those assumed at the beginning of this example problem. The values used at the beginning are based on the assumption that the steel below the neutral axis and the concrete above are both fully stressed. This is not actually the case because when the precast beam is cast in the casting yard, all the steel is placed in the beam. This amount of steel must also help to carry the composite live load bending moment, so that the actual steel stresses in the precast section under dead load only should be low.

Dead load steel stress

$$f_s = \frac{M_D}{A_s jd} = \frac{36(12{,}000)}{(3.0)(0.86)(16.56)} = 10{,}110 \text{ psi}$$

Dead load concrete stress

$$f_c = \frac{f_s}{n}\left(\frac{kd}{d - kd}\right) = \frac{10{,}110}{8}\left(\frac{6.83}{9.73}\right) = 887 \text{ psi}$$

Check temporary construction stresses.
Total moment under dead load and construction load is

$$36 + 21.6 = 57.6 \text{ kip-ft}$$

$$f_s = 10{,}110\left(\frac{57.6}{36}\right) = 16{,}180 \text{ psi}$$

$$f_c = 887\left(\frac{57.6}{36}\right) = 1{,}420 \text{ psi} < 1{,}800 \text{ psi}$$

Live Load on the Composite Member:

First locate the neutral axis of the composite section. Is the neutral axis in the slab or in the stem of the member? The neutral axis should be somewhere close to the junction of the slab and the stem. To determine the location, take moments of areas about the junction of slab and stem.

Since we are now dealing with the slab concrete, use $n = 9$. Use the slab concrete and transform the steel reinforcing into an equivalent area of concrete.

$$\text{SLAB} = \text{beam steel (transformed)}$$

$$(b)(t)\left(\frac{t}{2}\right) = nA_s(16.56)$$

$$(72)(4)\left(\frac{4}{2}\right) = (9)(3)(16.56)$$

$$576 > 447$$

The slab factor is larger, so the neutral axis is in the slab.

$$(0.72)(x)\left(\frac{x}{2}\right) = (9)(3)(20.56 - x)$$

from which $x = 3.56$ in.

Live Load Stresses on the Composite Section

For the composite beam,

$$jd = d - \frac{x}{3} = 20.56 - \frac{3.56}{3} = 19.38 \text{ in.}$$

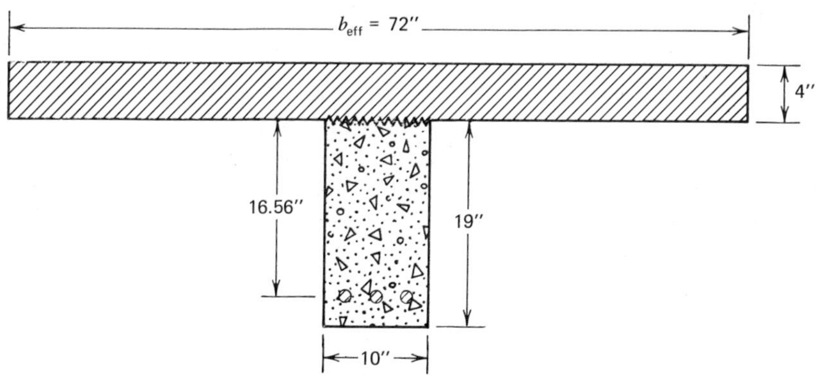

FIGURE 4.8 Composite T beam.

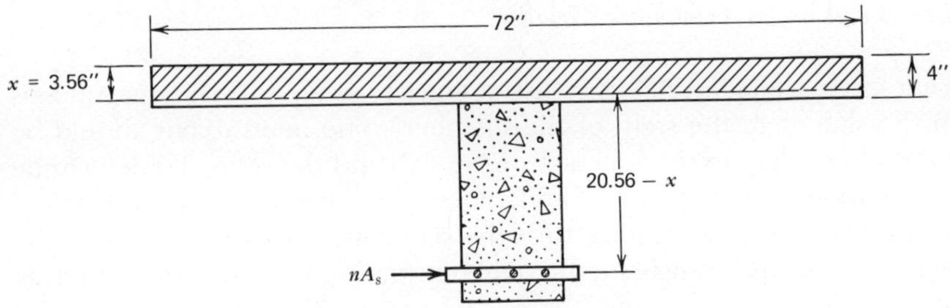

FIGURE 4.9 T beam showing transformed steel.

Live load steel stress

$$f_s = \frac{M}{A_s jd} = \frac{(64.8)(12{,}000)}{(3.0)(1{,}938)} = 13{,}400 \text{ psi}$$

Live load concrete stress

Since the neutral axis of the composite member lies in the slab, there is no additional live load concrete stress at the top of the precast member. Check concrete stress at top of slab:

$$f_c = \frac{f_s}{n}\left(\frac{x}{d-x}\right) = \frac{13{,}400}{9}\left(\frac{3.56}{17.0}\right) = 312 \text{ psi}$$

TOTAL BENDING STRESSES—LIVE LOAD + DEAD LOAD

$$\text{STEEL} = 13{,}400 + 10{,}110 = 23{,}510 \text{ psi} < 24{,}000 \text{ psi}$$

Concrete

TOP OF BEAM	1420 psi
TOP OF SLAB	312 psi

Shear Connection

The alternate design method of the ACI code requires that allowable stresses for shear in beams be reduced to 55% of the values used with the strength method. However, no ϕ or overload factors are applied to the shear loads, so that the net result is virtually no difference between the methods.

$$V_D = \frac{0.5(24)}{2} = 6 \text{ kips}$$

$$V_L = \frac{0.9(24)}{2} = 10.8 \text{ kips}$$

TOTAL $\quad V_L = 16.8 \text{ kips}$

$$\text{ALLOWABLE STRESS} = 0.55 \times 80 \text{ psi} = 44 \text{ psi}$$

$$v = \frac{V}{bd} = \frac{16{,}800}{(10)(20.56)} = 82 \text{ psi} > 44 \text{ psi}$$

Consequently, both extended stirrups and intentional roughening are required. The web steel requirement can be determined as in the previous example.

4.5 Construction Loads

Construction loads can be especially critical in concrete members. This is particularly true if the beam member is cast in place. The modulus of elasticity of concrete changes in the very early stages of its life. The green concrete may have enough structural integrity to support the load, but it may deflect enough to permanently damage the member.

According to ACI, the ultimate strength of shored and unshored members is the same. However, Section 9.5 of the ACI code does provide for different treatment of the deflection of shored and unshored members.

4.6 Stirrup Configurations

In the event that the closed-loop stirrup does not provide enough embedment length, some unusual configurations of the extended stirrups may be required (see Fig. 4.10).

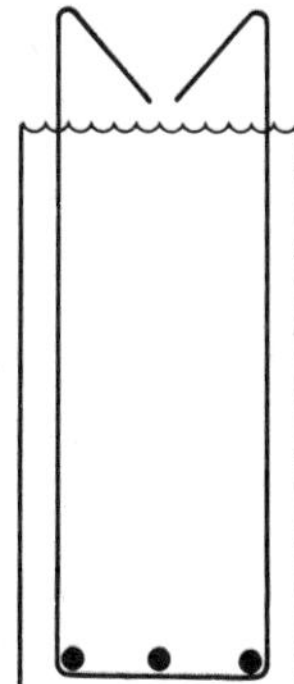

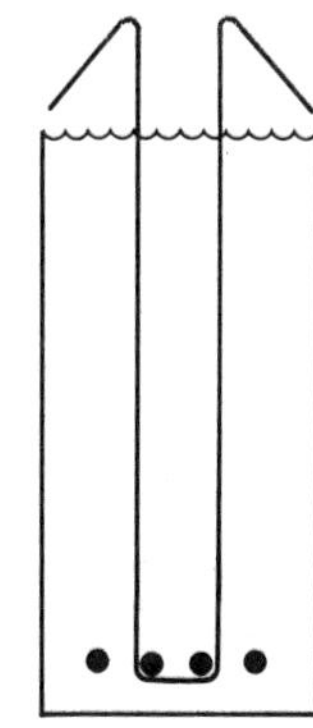

FIGURE 4.10 Stirrup configurations.

4.7 Prestressed Concrete

Concrete-concrete composite members can also be built using prestressed concrete. The logical combination is to use precast, prestressed members with a cast-in-place deck. The ACI code makes no distinction in the shear connection between precast, prestressed members and precast members with ordinary reinforcement. The code also recognizes that the requirements for the effective width of slab in T beam elements may be overly restrictive for prestressed concrete. The code recommends the use of the normal effective width requirements, but it permits the final choice to be based on the experience and judgment of the engineer.

However, in the composite beam using a precast, prestressed beam and a cast-in-place nonprestressed slab, the requirements for the effective width of slab must be used. The precast, prestressed beam member should be designed in accordance with Chapter 18 of ACI Code 318-71.

In bridge work, Section 1.6 of the AASHTO specifications covers the design of prestressed members. The requirements parallel those for other concrete-concrete composite members. All the web reinforcement in the precast beam must extend into the cast-in-place deck, and this web reinforcement can be used to satisfy the horizontal shear requirement. The minimum total area of ties per foot of span shall not be less than the area of two No. 3 bars at 12-in. spacing.

Full transfer of the horizontal shear force can be assumed if the top of the precast beam is intentionally roughened, the precast beam is designed to resist the entire vertical shear, and the vertical ties are fully anchored into the slab. The closed-loop stirrup is the best way to anchor the beam and slab together. If these requirements are not all met, the horizontal shear must be calculated. AASHTO uses the static formula $v = V_uQ/Ib$ rather than the strength design formula $v = V/\phi bd$ for computing the horizontal shear.

Allowable stresses for shear are as follows:

1. When the minimum tie requirements are met—75 psi.
2. When the minimum tie requirements are met and surfaces are intentionally roughened—300 psi.
3. In addition to the above, for each percent of vertical reinforcement crossing the joint, in excess of the minimum fire requirements—150 psi.

Because of the cyclic nature of bridge loadings, AASHTO requires a positive tie-down and does not permit intentional roughening alone as a satisfactory shear connection.

When the slab is cast onto the precast beams, the slab bonds to the beam and to the vertical reinforcement, and differential shrinkage stresses are set up in the slab and in the bottom of the beams. This shrinkage effect should be investigated, and if necessary, added to the design loads.

References

4.1 Grossfield, B., and Brinstiel, C., "Tests of T-Beams with Precast Webs and Cast-in-Place Flanges," *ACI Journal Proceedings,* Vol. 59, No. 6, June 1962.

4.2 Mattock, A. H., and Kaar, P. H., "Precast Prestressed Concrete Bridges: (4) Shear Tests of Continuous Girders," *Journal PCA Research and Development Laboratories,* Vol. 3, No. 1, January 1961.

4.3 Saemann, J. C., and Washa, G. W., "Horizontal Shear Connections Between Precast Beams and Cast-in-Place Slabs," *ACI Journal Proceedings,* Vol. 61, No. 11, November 1964.

4.4 Hanson, N. W., "Precast-Prestressed Concrete Bridges, (2) Horizontal Shear Connections," *Journal PCA Research and Development Laboratories,* Vol. 2, No. 2, May 1960.

4.5 Wang, C. K., and Salmon, C. G., *Reinforced Concrete Design,* Intext Publishers, New York, 1973.

4.6 Lin, T. Y., "Design of Prestressed Concrete Structures", John Wiley & Sons, Inc., N.Y. 1972.

5

Timber-Steel Construction

Wood and steel have long been used together in structures, but they have not been given the attention commanded by other types of composite construction. Wood and steel can be used in a flitched beam or in a coverplated beam.

Wood and steel were used together over a century ago in the old Howe and Pratt trusses and are still being used in some modern truss structures. More recently, the wood-steel combination has entered the market as an open web joist.

5.1 The Coverplated Timber Beam

The use of a steel coverplate can almost double the carrying capacity of the beam. However, in new construction where head room is no problem, the coverplated beam is usually not the most economical solution. A slightly deeper wood beam alone has the same carrying capacity as the coverplated beam and does the job for less cost.

One potential use for the coverplated beam is in rehabilitation work. The load carrying capacity of an existing heavy timber structure can be increased by adding the steel coverplate. This can be done without having to remove the existing floor system.

The coverplate of the proper width should have the holes drilled for the lag bolt shear connectors. This drilling (or punching) should preferably be done by the steel fabricator before the plate is delivered to the site. If this is not possible, the holes should be drilled while the plate is on the ground, before it is lifted up into position. If the timber beam is continuous, the beam will have to be temporarily jacked up off its interior support so that the steel plate can be positioned. Holes for the lag bolt connectors can easily be drilled into the wood beam, once the steel plate is in position.

5.1.1 Example

The problem is an existing, heavy timber, milltype building. The floor beams are nominal 6 × 12-in. timbers on 5-ft centers. The existing floor system was designed for a load of 60 psf. Management wishes to utilize the present structure and upgrade the floor system to carry 100 psf. The beam span is 20 ft.

The solution used is the addition of a $\frac{3}{8}$-in. steel plate to the bottom of the existing beam.

Properties

WOOD	Allowable bending stress = 1,600 psi
	Allowable shear stress = 100 psi
	Modulus of elasticity = 1,600,000 psi
STEEL	Allowable stress = 21,600 psi
	Modulus of elasticity = 30,000,000 psi
MODULAR RATIO	$n = \dfrac{E_s}{E_w} = \dfrac{30 \times 10^6}{1.6 \times 10^6} = 18.75$

Check the capacity of the wood section alone:

Section Modulus

$$Z = \frac{bd^2}{6} = \frac{(5.5)(11.5)^2}{6} = 121.2 \text{ in.}^3$$

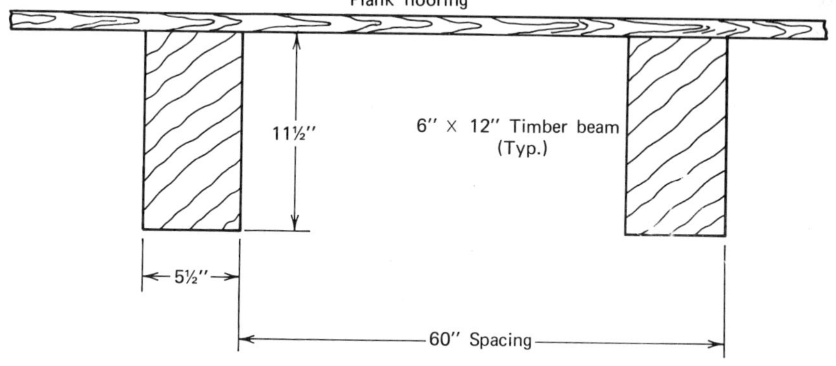

FIGURE 5.1 Existing beam.

Resisting Moment

$$M_r = fZ = 1{,}600 \times 121.2 = 193{,}970 \text{ in.-lb}$$

Using the 20-ft span,

$$M = \frac{wL^2}{8} \times 12$$

$$w = \frac{8M}{12L^2} = \frac{8(193{,}970)}{12(20)^2} = 323 \text{ ppf}$$

CAPACITY OF EXISTING BEAM $\quad = \dfrac{323}{5} = 64.6 \text{ psf}$

Now add the steel plate.

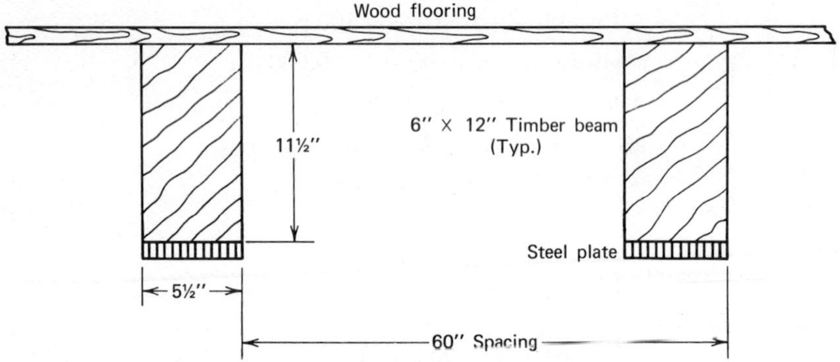

FIGURE 5.2 Plated beam.

Locate the neutral axis. Transform the steel into an equivalent amount of wood. Take moments of areas about the base of the section shown in Figure 5.2.

		A	y	A_y
PLATE	$\frac{3}{8}(5.5)(18.75) =$	38.67	$\times 3/16 =$	7.25
BEAM	$(5.5)(11.5) =$	63.25	$\times 6.13 =$	382.40
		101.92		394.65

$$\bar{y} = \frac{394.65}{101.92} = 3.87 \text{ in.}$$

Determine moment of inertia (refer to Fig. 5.2):

TOP WOOD $\dfrac{5.5(8.0)^3}{3} = 938.7$

BOTTOM WOOD $\dfrac{5.5(3.5)^3}{3} = 78.6$

STEEL PLATE $(\tfrac{3}{8})(5.5)(18.75)(3.68)^2 = 523.7$

1541.0 in.^4

In the calculation of I, we have ignored the moment of inertia of the plate about its own center of gravity because this term is negligible when compared to the other terms.

Carrying Capacity

TOP WOOD FIBERS $M_r = \dfrac{fI}{c} = \dfrac{1{,}600 \times 1{,}541}{8.0} = 308{,}200 \text{ in.-lb}$

BOTTOM WOOD FIBERS $M_r = \dfrac{fI}{c} = \dfrac{1{,}600 \times 1{,}541}{3.5} = 704{,}460 \text{ in.-lb}$

BOTTOM OF STEEL PLATE $M_r = \dfrac{fI}{c} = \dfrac{1{,}600 \times 1{,}541}{3.87} = 637{,}100 \text{ in.-lb}$

The top wood fibers govern the design.

$$w = \frac{8M}{12L^2} = \frac{8(308{,}200)}{12(20)^2} = 514 \text{ ppf}$$

CARRYING CAPACITY IN psf $\dfrac{514}{5} = 102.8 \text{ psf}$ OK

Check the stress in the steel plate:

$$f = \frac{Mc}{I} \times 18.75 = \frac{308{,}200(3.87)}{1{,}541}(18.75) = 14{,}512 \text{ psi}$$

The Shear Connection

HORIZONTAL SHEAR IN WOOD

$$v = \frac{VQ}{Ib}$$

$$v = \frac{(514 \times 10)(5.5 \times 8 \times 4)}{1541 \times 5.5} = 107 \text{ psi} < 110 \text{ psi}$$ OK

Shear Connectors

Try $\tfrac{3}{8} \times 8$-in. lag bolt connectors. The value of one lag bolt in single shear in this type of wood (from AITC) is 935 lb.

Spacing of Connectors

SHEAR TO BE CARRIED

$$\frac{VQ}{I} = \frac{(514 \times 10)[(3/8)(5.5)(18.75)(3.69)]}{1541} = 472 \text{ ppi}$$

SPACING
$$\frac{935}{472} = 2 \text{ in.}$$

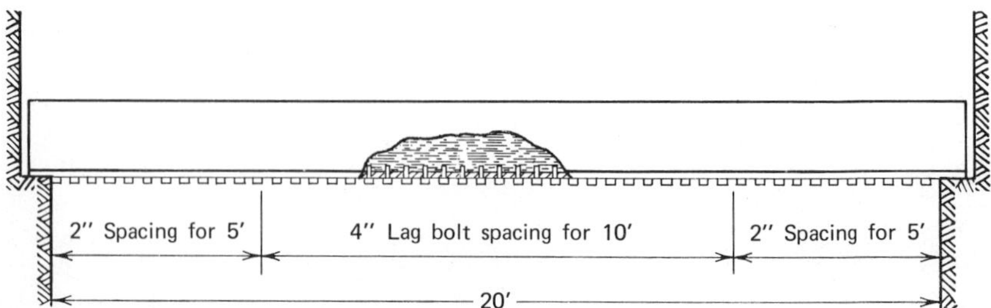

FIGURE 5.3 Lag bolt spacing. Spacings can be doubled by using the bolts in pairs.

This spacing can be used from the support to the quarter point of the beam and then increased.

The shear connection is one in which an epoxy adhesive could be used to great advantage. An epoxy adhesive has a shear strength of at least 1000 psi, which is more than adequate. A few lag bolts would be needed to keep the plate in place until the epoxy had cured. Unfortunately, most municipal building codes will not yet permit this type of connection.

5.2 The Flitch-Plate Beam

Another type of wood-steel composite member is made by placing a full depth steel plate between two wood joists. This type of beam, called a "flitched" beam, places the steel plate on edge and gives a high moment of inertia and section modulus. These members must be bolted tightly together to insure that they act as a unit. A thin steel plate on edge used as a beam would buckle laterally before developing anywhere near its full strength. The wood joists bolted to either side of the plate probably will provide enough lateral stiffness in the compression area of the beam.

Many of the usual construction methods will further stabilize this compression area. If wood joists are framed in on top of this girder, they can be toenailed in, providing a measure of lateral stiffness every 16 in.

A word of caution is in order about connections made to this flitched beam. There is usually only $1\frac{1}{2}$ in. of wood (the joist thickness) on either side of the steel plate. The ordinary 6-penny nail is 2 in. long and the 8-penny nail is $2\frac{1}{2}$ in. long, so that connections to the flitched beam must be planned with some care, if they are to be nailed.

5.2.1 Example Problem

Compare the carrying capacity of two 2×10 wood joists spiked together against the capacity of the same two 2×10 joists with a $\frac{3}{8}$-in. steel plate between them.

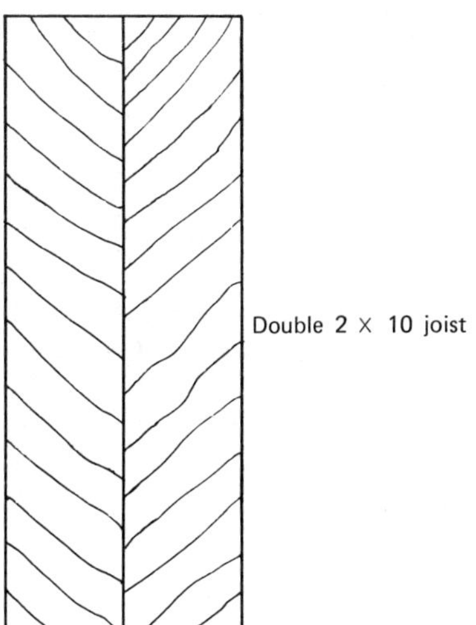

FIGURE 5.4

Capacity of the Wood Beam Alone

WIDTH $= 2 \times 1.5 = 3.0$ in. DEPTH 9.5 in.

Section Modulus

$$Z = \frac{bd^2}{6} = \frac{3(9.5)^2}{6} = 45.1 \text{ in.}^3$$

Using a wood with 1600 f:

Resisting Moment

$$M_r = fZ = 1{,}600 \times 45.1 = 72{,}200 \text{ in.-lb}$$

Now add the plate.

Figure 5.5 shows the actual beam and Figure 5.6 shows the beam with the steel plate converted into an equivalent area of wood.

Equivalent width of beam:

$$b = 2(1.5) + \tfrac{3}{8}(18.75) = 10.03 \qquad \text{Use 10 in.}$$

Section Modulus

$$Z = \frac{10(9.5)^2}{6} = 150.4 \text{ in.}^3$$

Resisting Moment

$$M_r = fZ = 1{,}600(150.4) = 240{,}666 \text{ in.-lb}$$

Adding the steel plate gives an increase of 230% in the resisting moment and carrying capacity of the beam.

5.3 Composite Steel-Wood Joists

Composite design and construction have also entered the open web joist field. The advantages of the open web joist have been known for many years. Among these advantages are

1. Long spans with minimum dead load.
2. Open webs provide space for ducts and wiring.
3. Easy handling and erection.
4. Designs and loads are well catalogued, so that a minimum of design time is required.
5. Fittings and bridging are readily available from the manufacturer.

The wood-steel joist offers the added advantage of having nailable chord members for the fast attachment of decking and ceiling materials. Figure 5.7 shows a composite joist. The joist is designed and fabricated as a pin-connected Warren truss. The joist shown in Figure 5.7 is one of the lighter joists, which uses 2 × 4 members that are oriented flatwise as the chords.

The composite joist is available in over a dozen profiles, including parallel chord, tapered, pitched, and curved chord. Figure 5.8 shows a

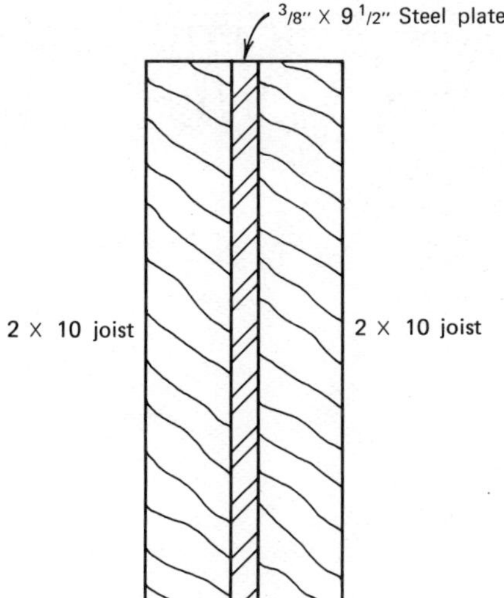

FIGURE 5.5 Flitched beam.

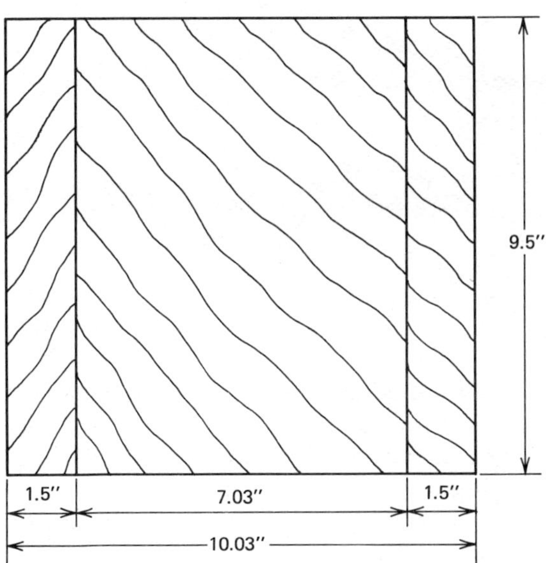

FIGURE 5.6 Transformed section of flitched beam.

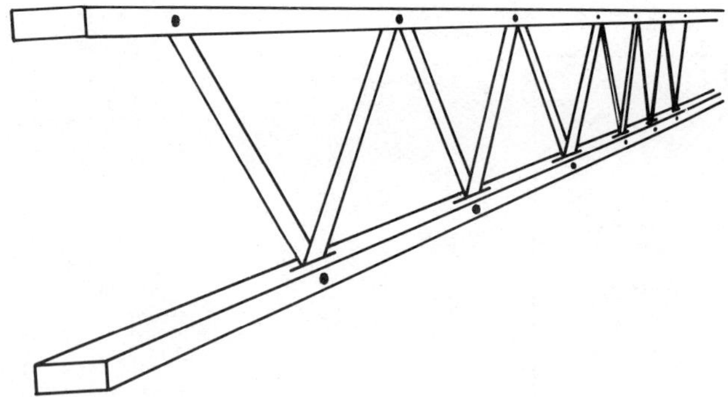

FIGURE 5.7 Composite bar joist.

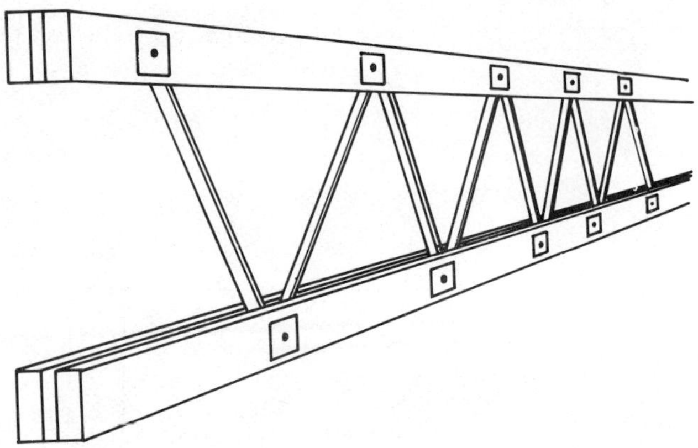

FIGURE 5.8 Composite bar joist.

heavier joist which uses two 2×4 members on edge as the chord members. For even heavier loads, this joist is available with double 2×6 chords.

As with any open web joist system, it must be remembered that these are lightweight structural components designed for vertical load. At the job site, they must be handled and erected with the same care as any other open web joist.

Figure 5.9 shows a parallel chord joist which was used on a small office building in Cincinnati, Ohio.

Figure 5.10 shows a longer span, pitched joist being used in a warehouse in Minneapolis.

FIGURE 5.9 Parallel chord joist. Small building in Cincinnati, Ohio. (Courtesy of Trus-Joist Corp, Boise, Idaho.)

FIGURE 5.10 Pitched joist. Minneapolis warehouse. (Courtesy of Trus-Joist Corp, Boise, Idaho.)

FIGURE 5.11 Construction of composite truss. Wesleyan hockey rink. (Courtesy of Warner, Burns, Toan, Lunde, Architects.)

FIGURE 5.12 Completed composite truss. Wesleyan hockey rink. (Courtesy of Warner, Burns, Toan, Lunde, Architects.)

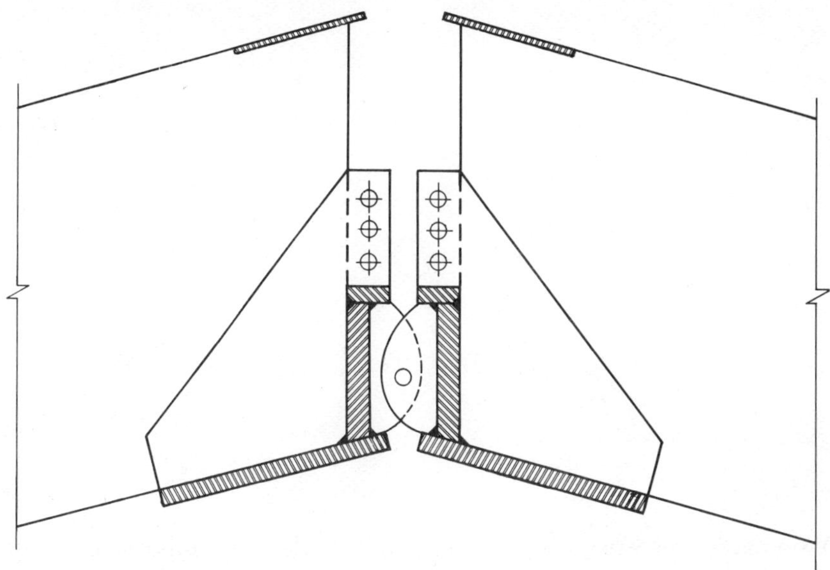

FIGURE 5.13 Hinge detail at the peak. (Courtesy of Warner, Burns, Toan, Lunde, Architects.)

5.4 The Composite Truss

The composite truss is as old as the Howe trusses of the past century and as new as the ultramodern hockey rink at Wesleyan University.*

This new composite truss uses a wooden top chord and high-strength steel cables for the lower chord. Web members and the diagonal bracing between trusses are of steel pipe. This truss provides a long, column-free span. The laminated wood top chord and purlins permit easy installation of the roof decking, and the slender webs and lower chord provide an uncluttered interior appearance. Figure 5.11 shows the truss nearly completed.

This is actually a very sophisticated composite design.† Figure 5.12, which is an interior view of the completed structure, shows that the building does not consist of a row of parallel trusses. The top chords of the trusses are pin-connected at the center (see Fig. 5.13). The 12 × 42-in. laminated top chord members for each half span are set diagonally so that they frame into an X connection at the center pin joint. The outer ends of the two top chord members then frame into each buttress.

In this particular structure, there was a lot of detailing required. However, once the detailing was completed, the construction was possible without too much interior shoring and bracing.

*Warner, Burns, Toan, Lunde, Architects.

†Severud-Perrone-Sturm-Conlin-Bandel, Structural Consultants.

6

Timber-Concrete Construction

Wood and concrete work especially well together in composite construction. Each has a great deal to contribute to the union. The beauty of natural wood, when left exposed, is unmatched by any other structural material. Wood is also our only replaceable natural resource, so that with good conservation practices, there will be a continuing supply of this material for many years. Wood also has an inherent flexibility, a capability to absorb large overloads of short duration without failure, so that most major specification writing bodies do not require the use of impact factors when designing and building in wood.

Concrete has good compressive strength and wearing characteristics. Concrete is also the only structural material that is in a plastic or moldable state when placed in the structure, so that it can be finished easily to line and grade to form an excellent deck.

Used together, these materials are a good combination. They have modulus values which are in the same order of magnitude (about 1.5 to 2.0×10^6 psi for wood and 2.0 to 3.0×10^6 psi for concrete). Because these values are quite close, the AASHTO specifications recommend a modular ratio of $E_c/E_w = 1.0$. This unit modular ratio greatly simplifies the design of the wood-concrete members.

Another factor which works for these materials when used together is the creep characteristics of both materials. In the steel-concrete composite member, creep and stress relaxation in the concrete will cause a shift in the neutral axis and consequently in the stress distribution under long-term loading. In the wood-concrete composite beam, both materials have creep and stress relaxation characteristics which are in the same order of magnitude, so that there should be very little change in the location of the neutral axis while the member remains in service.

One additional factor must be considered in the design and construction of the timber-concrete composite member. Additional shear connectors must be provided to take care of the additional horizontal shear stresses caused by expansion and contraction of the structure. In T-beam bridges, the wood is assumed to be unaffected by temperature changes, and therefore, the wood must resist the stress caused by expansion and contraction of the concrete. In a composite slab deck bridge, this extra shearing stress is neglected. However, joints should be provided in the concrete deck of the slab bridge. See Reference 6.4 for a method of handling these joints.

Timber-concrete composite construction usually falls into one of the two forms previously mentioned:

1. The T beam.
2. The timber-concrete deck.

6.1 The Timber-Concrete T Beam

This type of beam normally requires a large cross section when spans of a practical length are used. Because of the large sections, glued laminated members are frequently used for the beam stem. Because of the exposure conditions and the constant contact with the concrete, treated timber should be used. The T beam is commonly used for simply supported structures. There is no theoretical reason why this type of structure could not be used as a continuous beam. However, the practical construction problems lead us to the simple span solution. A continuous beam would either have to be fabricated in its full length or else field spliced. The longer continuous beam is, of course, more difficult to laminate, pressure treat, ship, handle, and erect. Also, the field splice might harm the natural line and beauty of the wood and could necessitate extra precautions in the wood treatment at the splice. AASHTO specifies treatment of bolt holes which are drilled at the site. This can be done, of course, but it is one more step in the field construction process, which costs money.

Practical and available sizes must be watched when ordering glulam members. The members are made up of either $\frac{3}{4}$ or $1\frac{1}{2}$-in. laminating stock. The usual board widths of $3\frac{1}{2}$, $5\frac{1}{2}$, $7\frac{1}{2}$, and so on, are used for lamination. However, the members are glued under pressure, so that there is a glue squeeze-out along the sides of the glulam member. Consequently, there is an extra step in dressing up the sides of the members, so that the width is further reduced. Standardization in the laminating industry has therefore resulted in widths of $2\frac{1}{4}$, $3\frac{1}{8}$, $5\frac{1}{8}$, $6\frac{3}{4}$, $8\frac{3}{4}$, $10\frac{3}{4}$, $12\frac{1}{4}$, and $14\frac{1}{4}$ in. These are the stock widths for glulam members. Other widths are available on special order from some laminators, but extra cost is involved.

As with any T beam, there are limitations on the effective width of the slab flange of the beam.

For bridges, these limitations according to AASHTO are as follows:

The effective flange width shall not exceed

1. One-fourth of the span length of the beam.
2. The distance center to center of beams.
3. Twelve times the least thickness of the slab.

For beams that have a flange on one side of the stem only, the effective flange width shall not exceed (see the exterior beam under the curb line in Fig. 6.1)

1. One-twelfth of the span length.
2. Six times the slab thickness.
3. One-half the center-to-center beam distance.

6.1.1 Forces and Stresses in the Beam and Slab

In the simple span structure, the wood beam stem takes the tension and the concrete deck takes the compressive force. In a bridge designed for H-20 loading, the bending moments are sizeable and the wood section should be quite deep. In composite action, the neutral axis naturally shifts upward, very close to the junction of the beam and slab, so that most of the wood is in tension and the entire slab in compression.

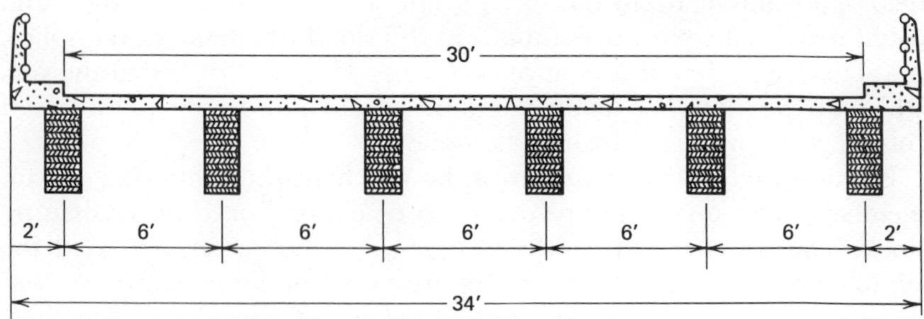

FIGURE 6.1 Cross section of a T-beam bridge.

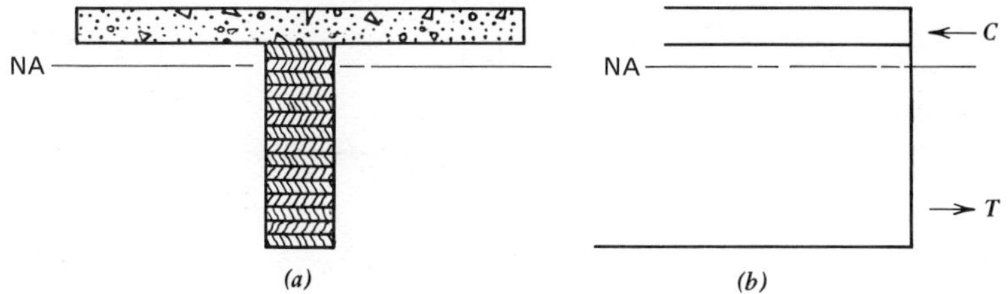

FIGURE 6.2 A glulam T beam.

Example 6.1 Design a composite timber-concrete T beam to span 32 ft and to carry AASHTO H-20 live load. Center-to-center spacing of stringers is 6 ft.

CONCRETE	$f'_c = 3000$ psi	$f_c = 1200$ psi	$E_c = 3.0 \times 10^6$ psi
WOOD	$F_b = 1600$ psi	$F_v = 145$ psi	$E_w = 1.7 \times 10^6$ psi

Estimate Beam Size and Size Thickness

AASHTO load tables give a bending moment of 266.5 kip-ft for a 32-ft span. This is loading for one lane. In our structure, two beams will carry the full lane width, so use one-half of this moment value for design. In this type of structure, it is reasonable to expect the live load to be approximately twice the dead load and the composite section modulus to be approximately twice the wood section modulus so that the live load and dead load stresses will be very roughly equal. Therefore, rough out a composite section modulus based on one-half of the allowable stress and pick out a wood section accordingly.

Composite Section Modulus Required for Live Load

$$S = \frac{M}{f} = \frac{(1/2)(266.5)(12{,}000)}{1/2 \times 1{,}600} = 2{,}000 \text{ in.}^3$$

The wood section modulus should be about one-half of this value. Since we are using a very deep laminated member, AITC specifications require the use of a size factor. Estimate this size factor at 0.8 and pick a wood member whose section modulus equals $1000 \div 0.8 = 1250$ in.^{3} From Table 2.16 of the AITC *Timber Construction Manual*, try the following member.

	$8\frac{3}{4} \times 30$ in.
SIZE FACTOR, ACTUAL	0.9
SECTION MODULUS	1,312.5 in.^{3}
	$I = 19{,}687$ in.^{4}
WEIGHT PER FOOT	$1.82 \times 30 = 54.6$ ppf

Use a 6-in.-thick concrete slab.

Determine Effective Flange Width

$$
\begin{aligned}
1/4 \text{ span} &= 1/4(32)(12) = 96 \text{ in.} \\
c/c \text{ spacing} &= 6(12) \qquad = 72 \text{ in.} \\
12 \times \text{slab } t &= 12(6) \qquad = 72 \text{ in.} \longleftarrow \text{controls}
\end{aligned}
$$

Since AASHTO specifies $E_c/E_w = 1$, the effective width of the transformed slab area also equals 72 in.

Determine the location of the neutral axis and find the transformed moment of inertia. Take moments of areas about the bottom of the beam:

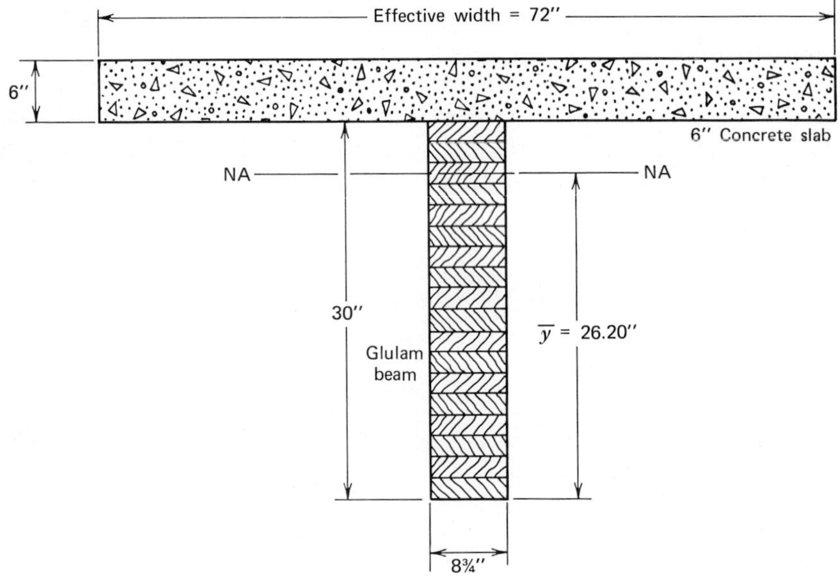

FIGURE 6.3

	A	y	Ay
WOOD STEM	$(8.75\times 30)=262.5$	$\times 15=$	3,937.5
SLAB	$(6\times 72)=432$	$\times 33=$	14,256
	694.5		18,193.5

$$\bar{y}=\frac{18{,}193.5}{694.5}=26.20 \text{ in.}$$

At this point, the figures are worth a second look because the neutral axis is still almost 4 in. below the slab. The 4 in. of wood above the neutral axis are in compression and are usually neglected in computing moment of inertia. However, quick checks of a few other practical sizes in this same area range also show a few inches of wood in compression. Adding this extra step to the computations would only increase the moment of inertia by 2 or 3%.

Moment of Inertia (ignore the wood in compression)

SLAB $$I_0=\frac{72\times(6)^3}{12}=1{,}296 \text{ in.}^4$$

$$Ad^2=(6\times 72)(6.80)^2=19{,}976 \text{ in.}^4$$

WOOD BEAM $$=\frac{(8.75)(26.2)^3}{3}=\frac{52{,}455}{73{,}727 \text{ in.}^4}$$

Now determine actual loads and bending moments:

Dead Load

SLAB $$\left(\frac{6}{12}\right)\left(\frac{72}{12}\right)(150)=450 \text{ ppf}$$

WOOD BEAM $$=\frac{54.6 \text{ ppf}}{504.6 \text{ ppf}}$$

$$M_D=\frac{504.6\times(32)^2(12)}{8}=839{,}000 \text{ in.-lb}$$

Dead Load Stress in Wood

$$f_b=\frac{839{,}000}{1{,}312.5}=639 \text{ psi}$$

Live Load Stresses

WOOD $$f_b=\frac{1{,}602{,}000(26.2)}{73{,}727}=569 \text{ psi}$$

CONCRETE $$f_c=\frac{1{,}602{,}000(9.8)(1.3 \text{ for impact})}{73{,}727}=277 \text{ psi}$$ OK

TOTAL WOOD STRESS $639+569=1{,}208$ psi

ALLOWABLE WOOD STRESS $1{,}600\times 0.9=1{,}440 \text{ psi}>1{,}208$ psi OK

This bridge beam is approximately 15% overdesigned and could be trimmed. However, leaving the bridge as is provides a margin for future repaving. If a paving allowance is included in the design, it should be added as a uniform load on the composite section because any repaving would not be done until the bridge was several years old.

6.1.2 Design and Installation of the Shear Connectors

The shear connection in the timber beam can be provided either by notching the top of the beam, or by lag screws or both. Notches in the top of the beam should be at least $\frac{1}{2}$ in. deep and preferably $\frac{3}{4}$ in. deep. The notches should be milled the full width of the beam. The AITC specifications state that if half the area of the top surface is notched, only half the area remains effective in shear resistance. Consequently, the allowable shear stress value for the wood is reduced by one-half. Since the wood should be notched before it is pressure treated, the number and size of the notches can be left as a practical matter for the fabricator to decide. If the fabricator has the equipment to mill a 2-in.-long notch the full width of the beam, then three notches per foot can be provided.

If lag bolts are to be used as shear connectors, they should be turned with a wrench, not hammered, into prebored holes. The hole diameter should be approximately two-thirds the nominal bolt diameter. Recommended values for various wood species can be found in the AITC *Timber Construction Manual.* For use as shear connectors, at least one-half the bolt length should be turned into the member. The extended portion of the bolt should have at least $1\frac{1}{2}$ in. of concrete cover. The bolt holes should be treated with preservative before installation of the lag bolts.

The best shear connection should not only resist horizontal shear, but it should also prevent the slab from lifting up and separating from the beam. In the event that notching the beam provides ample horizontal shear resistance, large spikes driven partially into the top of the member help to prevent uplift of the slab. If lag bolts are used, the heads of the lag bolts provide sufficient uplift resistance.

Check the Horizontal Shear Stress in the Example, Assuming a Notched Beam

$$F_v = \frac{1}{2}(145) = 72.5 \text{ psi}$$

$$F_v = \frac{VQ}{Ib}$$

where
$$V = 18{,}250 \text{ lb};$$
$$I = 73{,}727 \text{ in.}^4;$$
$$Q = (6 \times 72)(6.8) = 2{,}938 \text{ in.}^3;$$
$$b = 8.75.$$

$$F_v = \frac{(18{,}250)(2{,}938)}{(73{,}727)(8.75)} = 83.1 \text{ psi} > 72.5 \text{ psi}$$

∴ Additional connectors are needed

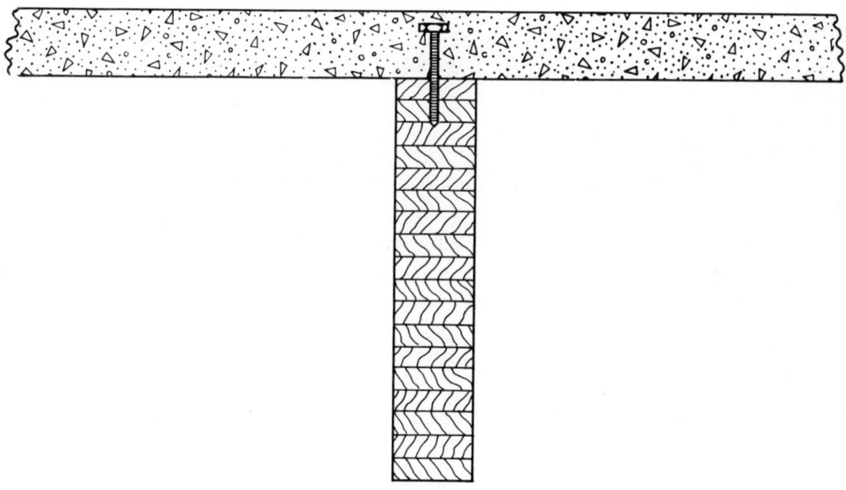

FIGURE 6.4

Use $\frac{3}{4} \times 8$-in. lag bolts.

The AITC *Timber Construction Manual* tabulates allowable values for lag bolts in lateral loading. Different tables take into account the type of member through which the load is applied. For instance, if two wood members are being joined, the values are different from those used if a stiff metal or concrete member is being connected to the wood. For our $\frac{3}{4} \times 8$-in. lag bolts, the allowable load per bolt (AITC Table 5.13) is 935 lb.

To determine the number of bolts required, simply multiply the excess amount of shearing stress by the top surface area of the beam. This gives the total force to be carried by the connectors.

$$N = \frac{(84.5 - 72.5)(8.75 \times 32 \times 12)}{935} = 43 \text{ bolts}$$

Determine the number of temperature connectors. Assume the temperature differential between beam and slab is 60F.

Concrete Stress

$$f_c = (0.0000065)(60)(3 \times 10^6) = 1170 \text{ psi}$$

(Note that this concrete stress is more than five times the live load stress in the concrete.)

Only one-third of the concrete flange area is assumed to be restrained by the timber stem, so use this factor in computing the number of temperature change connectors.

$$N = \frac{(1/3)(6 \times 72)(1170)}{935} = 180 \text{ bolts}$$

The total number of connectors required = 43 + 180 = 223 bolts throughout the span. Using the bolts in pairs, a uniform spacing of $3\frac{1}{4}$ in. between the pairs will provide 220 bolts, which should be sufficient.

Once again, if notching the beam provides ample shear resistance, large spikes can be partially driven into the top of the member to help prevent slab uplift. These spikes are generally inclined slightly toward the far support.

6.2 Timber-Concrete Composite Slab Bridges

The timber-concrete deck bridge has some construction advantages over the T-beam type. In the T-beam type, the slab must still be formed. In building construction, it is relatively simple to shore up the beam and slab forms. In bridge work, it is seldom feasible to construct an extensive system of falsework because of the practical difficulty of providing a solid base for so many temporary supports.

The slab bridge consists of planks, set on edge and spiked, or doweled together. Alternate sizes of planks, such as 2×4 with 2×6 and 2×8, form the wood section shown in Figure 6.5. This type of deck is generally used for short spans, so that the individual wood pieces can be handled by one or two men, with a minimum amount of help from hoisting equipment. Once the wood deck is placed, it provides a platform for the workmen and also acts as the bottom form for the concrete.

This type of structure lends itself well to a series of short spans. It provides a good-looking bridge because of the slender thickness of the deck and the solid appearance when seen from below.

The riding surface of the roadway should be crowned. This crown can be provided either by additional concrete toward the centerline of the

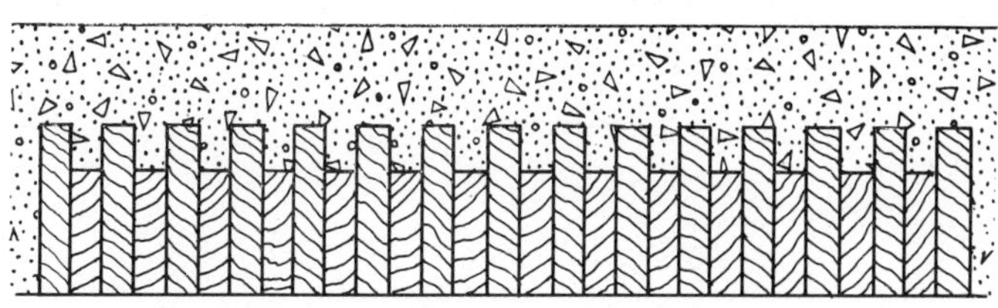

FIGURE 6.5 Timber-concrete slab deck.

roadway or by placing risers on the pier caps which will give a concrete mat of constant thickness.

Since this type of deck is usually used for a series of continuous spans, the laminations have to be furnished in practical lengths and butt jointed. One-third of the laminations are joined over the supports, and the remaining two-thirds are placed at alternate quarter points, where the dead load bending moment is close to zero. The reason for joining some of the members over the supports, of course, is to make the structure easier to build.

The reinforcing steel in the deck provides shrinkage and temperature reinforcement, and also provides for the negative bending moments over the interior supports. The steel should be continuous over the supports. Splices in the steel should be made in the center positive moment region of the bridge where the concrete is normally in compression.

6.2.1 Example Problem

Design and construct a continuous timber-concrete slab deck highway bridge. There are two 24-ft spans. The roadway width has two 12-ft-wide lanes with a 3-ft shoulder on each side. The total roadway width, face to face of the curb, is 30 ft.

Allowable Stresses

WOOD,	BENDING	$F_b = 1800$ psi
	SHEAR	$F_v = 145$ psi
	CONCRETE	$f'_c = 3{,}000$ psi
	STEEL	$f_s = 20{,}000$ psi
	DEAD LOAD	50 pcf for treated wood; 150 pcf for concrete

Rough Out a Deck Size

Assume a deck to be 12 in. thick. Alternating 2×6 and 2×8 planks gives an average thickness of $6\frac{1}{2}$ in. of wood. This means an average of $5\frac{1}{2}$ in. of concrete.

Estimated Dead Load

WOOD $\quad \dfrac{6.5}{12} \times \dfrac{12}{12} \times 50 = 27.1 \text{ ppf}$

CONCRETE $\quad \dfrac{5.5}{12} \times \dfrac{12}{12} \times 150 = \underline{68.7 \text{ ppf}}$

$95.8 \text{ ppf} \qquad$ say 96 ppf

Simple Beam Moment

$$M = \frac{wL^2}{8} = \frac{96(24)^2(12)}{8} = 82{,}944 \text{ in.-lb}$$

Section Modulus Required

$$S = \frac{82{,}944}{1{,}800} = 46.1 \text{ in.}^3$$

Section Modulus Supplied

$$S = \frac{12(6.5)^2}{6} = 84.5 \text{ in.}^3 > 46.1 \text{ in.}^3$$

The estimated size looks fine.

This is the section modulus of the wood deck alone, which must support all the dead load until the concrete has cured enough to cause the composite action.

The wood deck at this stage looks overdesigned. However, the construction sequence may govern the design. There is the possibility that the dead load deflection of the deck under the weight of the wet concrete may not be acceptable.

Determine Properties of the Wood Deck

Determine location of the neutral axis:

	A	y	Ay
$(1.5)(7.5) =$	$11.25 \times$	$\dfrac{7.5}{2}$	$= 42.19$
$(1.5)(5.5) =$	$\underline{8.25} \times$	$\dfrac{5.5}{2}$	$= \underline{22.69}$
	19.50		64.88

$$y = \frac{64.88}{19.50} = 3.33 \text{ in.}$$

Moment of Inertia of the Wood Section at Midspan

In a 1-ft-wide section of the deck, there are 6 in. of width of 2×6 planks and 6 in. of width of 2×8 planks.

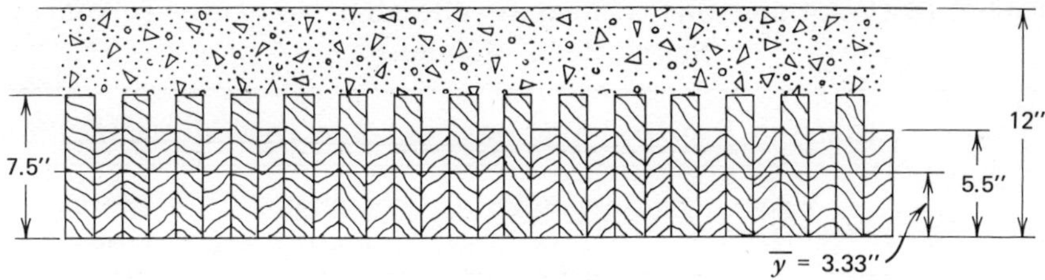

FIGURE 6.6

I *Wood*

SECTION BELOW NEUTRAL AXIS $\frac{12 \times (3.33)^3}{3} = 147.7$

SECTION OF 2×6 ABOVE NEUTRAL AXIS $\frac{6 \times (2.17)^3}{3} = 20.4$

SECTION OF 2×8 ABOVE NEUTRAL AXIS $\frac{6 \times (4.17)^3}{3} = 145.0$

313.1 in.^4

Since one-third of the laminations will be joined over the interior support, only two-thirds of the wood section is considered effective in resisting the negative bending at the support.

AT INTERIOR SUPPORT $\frac{2}{3} I_w = \frac{2}{3}(313.1) = 208.7 \text{ in.}^4$

Table 6.1 below gives the percentages of simple beam moments which are specified by AASHTO for the design of a continuous slab-type deck.

Actual Dead Load

WOOD 2×6 $\left(\frac{6}{12} \times \frac{5.5}{12}\right) 50 = 11.5 \text{ ppf}$

2×8 $\left(\frac{6}{12} \times \frac{7.5}{12}\right) 50 = 15.6 \text{ ppf}$

CONCRETE $\left(\frac{6}{12} \times \frac{4.5}{12}\right) 150 = 28.1 \text{ ppf}$

$\left(\frac{6}{12} \times \frac{6.5}{12}\right) 150 = 40.6 \text{ ppf}$

95.8 ppf

TOTAL DL = 95.8 ppf Use 96 ppf

Simple Beam Moments—Dead Load

$$M = \frac{wL^2}{8} = \frac{96(24)^2}{8}(12) = 82{,}944 \text{ in.-lb}$$

Table 6.1 Continuous Spans (Maximum Bending Moments—Percent of Simple Beam Moments[a])

	Uniform Dead Load Moments				Live Load Moments			
	Wood Subdeck		Composite Slab		Concentrated Load		Uniform Load	
	Pos	Neg	Pos	Neg	Pos	Neg	Pos	Neg
Span	Percentage of Simple Beam Bending Moments							
Interior	50	50	55	45	75	25	75	55
End	70	60	70	60	85	30	85	65
Two-Span	65	70	60	75	85	30	80	75

[a] From *Standard Specifications for Highway Bridges*, adopted by AASHTO, 11th ed. 1973.

Design Moments for Continuous Spans (from Table 6.1)

POSITIVE (MIDSPAN) $= (0.65)(82{,}944) = 53{,}914$ in.-lb

NEGATIVE (SUPPORT) $= (0.60)(82{,}944) = 58{,}000$ in.-lb

Bending Stresses in Wood—Dead Load

MIDSPAN $$F_b = \frac{M_d}{I_w} = \frac{53{,}914(3.33)}{313.1} = 573 \text{ psi}$$

SUPPORT $$F_b = \frac{M_d}{I_w} = \frac{(58{,}060)(4.17)}{208.7} = 1{,}160 \text{ psi}$$

The stresses look satisfactory. When the live load stresses are added, the total stresses should be below the allowable levels.

However, now check the deflection of the wood deck under dead load. Using the deflection coefficient from a structural handbook for maximum deflection in a two-span structure;

DEFLECTION $$= 0.0092\,\frac{wL^4}{EI}$$

$$= \frac{0.0092(96)(24)^4(1{,}728)}{(1{,}600{,}000)(313.3)} = 1.0 \text{ in.}$$

The wood deck will deflect 1 in. under the full dead load. If the project engineer has set fixed bench marks for the screeding of the slab, another inch of concrete will be required to bring up the slab to line and grade.

However, this 1 in. of concrete is another 12.5 lb of dead load which causes further deflection! If the riding surface is to be crowned by adding additional concrete at the highway centerline, the problem is further compounded. This crowning would add another $1\frac{1}{2}$ in. of concrete at the centerline, tapering off to zero at the curb line.

Since deflection varies as the fourth power of span length, it is easier to control the span than the load. One temporary bent at midspan, effectively cutting the bridge into 12-ft spans during construction, would solve the problem.

If the governing specifications call for an additional 50 psf of temporary construction load, the shorter spans would still handle the problem quite effectively.

Live Loads

The live loads act on the composite section after the concrete has cured. The load on one rear wheel of an H-20 truck is 16,000 lb. This is distributed over 5 ft of width.

$$P = \frac{16{,}000}{5} = 3{,}200 \text{ lb}$$

If this rear wheel of the standard AASHTO truck is at midspan, the front wheel will be very close to the support in the adjacent span where it will have little effect on the positive bending moment.

$$\text{SIMPLE BEAM MOMENT} \quad = \frac{PL}{4} = \frac{3{,}200(24)(12)}{4} = 230{,}400 \text{ in.-lb}$$

Design Moments for Continuous Spans

$$+(0.85)(230{,}400) = 195{,}840 \text{ in.-lb}$$
$$-(0.30)(230{,}400) = 69{,}120 \text{ in.-lb}$$

Properties of the Composite Section

The assumed deck thickness of 12 in. with alternating 2 × 6 and 2 × 8 planks gives a minimum concrete thickness of $4\frac{1}{2}$ in. above the high laminations. This is enough thickness to provide satisfactory cover for the reinforcing steel. Use $E_c/E_w = 1$ and $E_s/E_w = 18.75$. These are both in accordance with AASHTO specifications.

In the positive moment region, since $E_c/E_w = 1$, the 12-in. deck is assumed to be a homogeneous section:

$$I = \frac{12(12)^3}{12} = 1728 \text{ in.}^3$$

In the negative moment regions, compute the shrinkage and temperature steel requirements first. These bars should be relatively small at a fairly wide spacing. If the hoisting equipment is available, these mats can be fabricated on the ground in relatively large sections and hoisted into position. Then additional bars can be added over the interior supports to make up the necessary negative moment steel.

Shrinkage and Temperature Steel

$$p = 0.0020 \qquad A_s' = 0.0020(12)(5.5) = 0.132 \text{ in.}^2/\text{ft}$$

Number 4 bars at 8 in. provide 0.29 in.^{2}/ft. This steel size and spacing will be used in both directions. Additional No. 4 bars at 8 in. can be placed longitudinally over the support. In this way, the total steel over the support is No. 4 bars at 4-in. spacing.

NEGATIVE $\qquad A_s = 0.59 \text{ in.}^2/\text{ft}$

TRANSFORMED STEEL AREA $\qquad 0.59(18.75) = 11.06 \text{ in.}^2/\text{ft}$

Determine the location of the neutral axis for the negative section at support.

	A	y	Ay
WOOD	$\frac{2}{3}(6)(5.5) = 22$	$\times \left(\frac{5.5}{2}\right)$	$= 60.5$
	$\frac{2}{3}(6)(7.5) = 30$	$\times \left(\frac{7.5}{2}\right)$	$= 112.5$
STEEL	$11.06 \times$	9.5	$= 105.1$
	63.06		278.1

$$y = \frac{278.1}{63.06} = 4.4 \text{ in.}$$

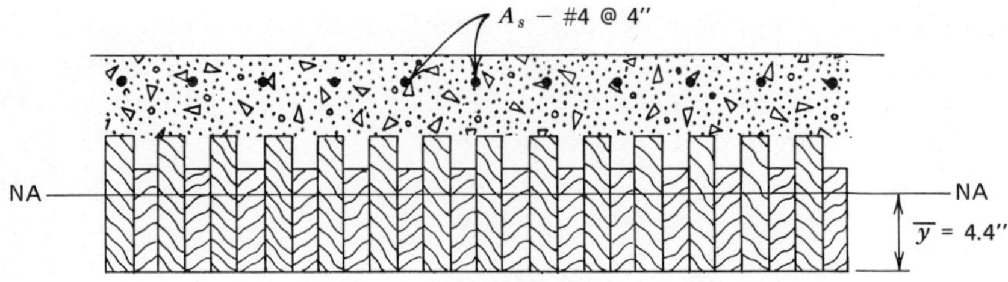

FIGURE 6.7

I, Negative

WOOD BELOW NEUTRAL AXIS $\dfrac{\frac{2}{3}(12)(4.4)^3}{3} = 227.2$

WOOD ABOVE NEUTRAL AXIS $\dfrac{\frac{2}{3}(6)(1.1)^3}{3} = 1.8$

WOOD ABOVE NEUTRAL AXIS $\dfrac{\frac{2}{3}(6)(3.1)^3}{3} = 39.7$

STEEL $(11.06)(5.1)^2 = \underline{287.7}$

556.4 in.^4

Bending Stresses: Live Load

MIDSPAN, WOOD $F_b = \dfrac{My}{I} = \dfrac{(195{,}840)(6)}{1{,}728} = 680 \text{ psi}$

CONCRETE $= 1.3\left(\dfrac{M_c}{I}\right)\left(\dfrac{E_c}{E_w}\right)$

$$= 1.3\,\frac{(195{,}840)(6)}{1{,}728}(1) = 884 \text{ psi}$$

Concrete stress is increased 30% for impact.

Bending Stresses: At Support

WOOD $F_b = \dfrac{M_c}{I} = \dfrac{(69{,}120)(3.1)}{556.4} = 385 \text{ psi}$

STEEL $f_s = 1.3\left(\dfrac{M_c}{I}\right)\left(\dfrac{E_s}{E_w}\right)$

$$= \frac{1.3(69{,}120)(5.1)}{556.4}(18.75) = 15{,}442 \text{ psi}$$

Table 6.2

Tabulated Values of Bending Stresses, psi

Material	Midspan	Support
Wood		
DL	573	1,160
LL	680	385
Total	1,253	1,545
Concrete	884	
Steel		15,442

6.2.2 The Shear Connection

The shear connection in the slab deck can be formed either by notching the wood members as in the T beam, or by using shear developer plates. The shear developer plates are trapezoidal-shaped plates about $\frac{3}{32}$ in. thick. They are driven into precut slots in the upper laminations of the deck. Slots for these shear plates should not be cut with a saw or with hand tools. The slots should be cut with a special chisel-blade, power tool which cuts the slot to the exact thickness and to the proper depth. Figure 6.8 shows a shear developer in place.

Capacity of the shear developers has been established by test. The plate shown in Figure 6.8 has a capacity of 1750 lb.

Whether shear plates or notches are used to develop the horizontal shear, 60*d* spikes are driven into the raised laminations to prevent uplift of the slab. The nails are not usually driven in vertically. They are inclined toward the support, and a little over an inch of the spike is left exposed to bond to the concrete. The spikes can be set on 2-ft centers with spacings alternating in adjacent laminations.

Placement of the Shear Developer Plates

For shear calculations, the heavy wheel load is considered to be distributed over a 4 ft width of the deck and is placed at three times the slab thickness from the support. This load position places both the front and

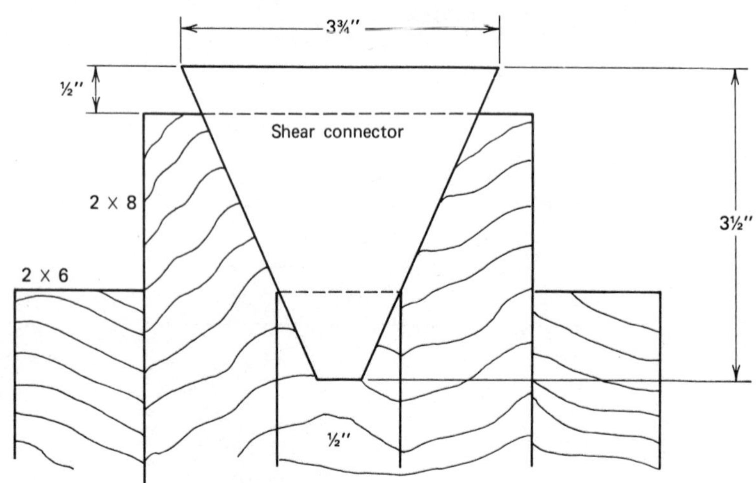

FIGURE 6.8 The shear developer plate.

rear wheels of the truck on the span. The load position is shown in Figure 6.9.

$$V = (7(1000) + 21(4000))\frac{1}{24} = 3792 \text{ lb}$$

Increase 30% for impact.

$$3792 \times 1.3 = 4929 \text{ lb}$$

$$f_v = \left(\frac{3}{2}\right)\left(\frac{V}{bd}\right) = \frac{3}{2}\left[\frac{4929}{(12)(12)}\right] = 51 \text{ psi at midspan}$$

At the support, shear is calculated only on the basis of those pieces which are continuous over the support, plus the transformed area of the steel. According to AITC, if the ends of the pieces are restricted, so that the effect of checking in the lumber is minimized, the allowable shearing stress value for the species may be doubled.

ALLOWABLE $F_v = 2(95) = 190$ psi

Shear at the Support

$$f_v = \left(\frac{3}{2}\right)\left(\frac{V}{bd}\right) = \left(\frac{3}{2}\right)\left(\frac{4929}{2/3(6)(5.5) + (2/3)(6)(7.5) + 11.06}\right) = 117.2 \text{ psi}$$

Both shear values, at midspan and support, are satisfactory with regard to the wood.

Developer Plates

Using $1\frac{1}{2}$-in.-wide lumber, there are four grooves in each foot of deck width.

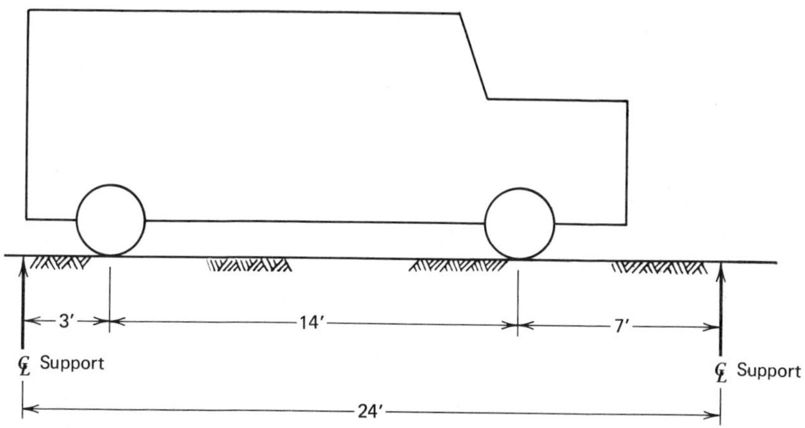

FIGURE 6.9 Truck load positioned for shear.

Shear in Each Groove Per Foot

AT SUPPORT $$\frac{117.2(144)}{4} = 4219 \text{ lb/ft}$$

AT MIDSPAN $$\frac{51(144)}{4} = 1836 \text{ lb/ft}$$

Each shear developer plate will account for 1750 lb.

SPACING, AT SUPPORT $$\frac{4219}{1750} = 2.4 \text{ plates per ft}$$

AT MIDSPAN $$\frac{1836}{1750} = 1 + \text{plates per ft}$$

Use a spacing of 5 in. from the support to the quarter point, then change to 12-in. spacing throughout the center one-half of the span. Stagger the plates in alternate grooves so that they do not form a continuous plane of weakness across the width of the slab.

6.2.3. Deflection Check

For live load deflection, use the heavy wheel load at midspan.

DEFLECTION $$= 1.3\left(\frac{PL^3}{48EI}\right) = \frac{(1.3)(3{,}300)(24)^3(1{,}728)}{48(1{,}600{,}000)(1{,}728)} = 0.75 \text{ in.}$$

This deflection is satisfactory for live load. AITC limits live load deflection in highway bridges to $L/300$. This gives an allowable deflection of about 1 in.

6.3 Concrete Quality

Most concrete specifications set a lower limit on concrete strength. According to the latest ACI specifications, a statistical method is used to determine the probability that most test results will fall *above* a certain minimum value.

However, as concrete strength goes up, the concrete generally becomes more brittle. The timber-concrete composite bridge is a relatively flexible structure. This type of bridge is better served by a lower strength, lower modulus concrete. This does not mean poor quality concrete. If anything, it demands better mix design to insure concrete in the 2500 to 3500 psi range. The concrete strength values should have an upper, as well as a lower, limit. If the ACI statistical method is used, it simply means extending the specifications to insure that the strength values of the concrete fall within a range of acceptable values.

6.4 Construction

Figures 6.10 and 6.11 are a cross section and a construction photo of the Tres Pinas Creek, Elkhorn Bridge in California. This bridge consists of three 30-ft spans. The roadway width is 24 ft clear, and the bridge is

FIGURE 6.10 Cross section. Tres Pinos Creek, Elkhorn bridge, San Benito County, California. (Courtesy of American Institute of Timber Construction.)

FIGURE 6.11 Erection photo. Tres Pinas Creek bridge. (Courtesy of American Institute of Timber Construction.)

designed for H-20 loading. The cross section shown in Figure 6.10 is composed of alternating 3 × 8 and 3 × 10 Douglas Fir, fastened with spiral dowels. Note that both the upper and lower laminations are notched. In both photos longitudinal grooves can be seen on the side of the high laminations. These grooves help to provide uplift resistance. A second look at the cross section shows two high laminations adjacent to each other. The construction photo shows this pattern repeating every nine laminations. Apparently these nine laminations were assembled on the ground and lifted into position. In the lower right-hand corner, a workman can be seen placing a temporary support approximately at midspan to control deflection until the slab has cured and there is composite action.

References

6.1 Pincus, G., "Behavior of Wood-Concrete Composite Beams," *J. Str. Div., ASCE,* Vol. 96, No. ST10, October 1970.

6.2 Pincus, G., "Bonded Wood-Concrete T-Beams," *J. Str. Div., ASCE,* Vol. 95, No. ST10, October 1969.

6.3 *Timber Construction Manual,* American Institute of Timber Construction, Wiley, New York, 1974.

6.4 Cook, J. P., *Construction Sealants and Adhesives,* Wiley, New York, 1970.

7

The Shear Connector

In the composite beam, the purpose of the shear connector is to tie the beam and slab together and force them to act as a unit. In order to accomplish this, the shear connector must do two things. It must resist the horizontal force that develops between the beam and slab as the composite member is loaded. Second, it should hold the slab down against the beam and prevent uplift of the slab as the unit is loaded.

If the beam and slab are allowed to slip relative to each other as the beam is loaded, there is little or no composite action. You can check this slip tendency simply by bending a deck of ordinary playing cards in your hand. As you bend the deck, the cards slide easily over one another because there is no shear resistance. If you clamp the ends of the deck tightly together with two hands, it takes a great deal more effort to bend the deck.

7.1 Development of the Connector

Paradoxically, the first shear connector was no connector at all. Various investigators had found that the natural bond between steel beams and concrete furnished an additional measure of strength to the member. Probably the first published work in this area was that of W. Basil Scott who compiled a set of load tables for "Steel Joists Embedded in Concrete" for a firm of British fabricators in 1911. By the time another dozen years had elapsed, several more studies had been made. In 1922, a series of tests at the National Physical Laboratory in Britain had been published by Redpath and Brown. About the same time, three other studies were under way. Caughey showed that composite action existed by testing bridge floors in Iowa. Three Canadian professors, McKay, Gillespie, and Leluau, were working on floor panels of the same type. In 1923, tests sponsored by the Truscon Steel Company at the University of Nebraska, Purdue, and the Massachusetts Institute of Technology probably used the first real shear connector. The beams in these tests had their flanges sheared and the sheared prongs bent upward into the concrete to help develop the composite action.

In 1929, Caughey and Scott recommended projecting bolt ends as shear connectors. By the early 1930s, the Swiss were using spiral shear connectors, and the timber industry in the United States was using shear developer plates for slab-type bridges.

Currently, the AISC manual lists allowable load values for both the channel and stud connectors, but the headed stud has clearly emerged as the most widely used shear connector today.

The headed stud carries the horizontal shear load in flexure, rather than by bearing of the concrete against the face of the connector. The head on the stud provides excellent uplift resistance. The lower end of the stud is filled with a fluxing agent. The stud is fitted into a special gun, and an electric arc is set up between the bottom of the stud and the beam flange so that a pool of molten metal is formed at the base of the stud. This pool is confined by a ceramic retainer in order to form a neat uniform weld. The stud gun pushes the stud shank into the molten metal to complete the operation. The studs before welding are slightly over length so that the stud after installation has the true length listed in the AISC manual.

The spiral shear connector, which formerly was widely used, is not mentioned in the latest editions of the AISC, AASHTO, or ACI specifications.

7.2 Types of Connectors

In steel-concrete composite members, many different types of connectors have been used. In the early days of composite construction, there were no members specifically designed as shear connectors that were readily available. As a result, engineers and fabricators used members that were available. Plates, angles, channels, Z sections, T's, small sections of I beams, square bars, and concrete reinforcing rods have all been used. As the spiral bar became more readily available, it gained in popularity. Today, however, the two most frequently used connectors are the channel and the welded stud. Figure 7.1 shows several of the types of connectors which have been used.

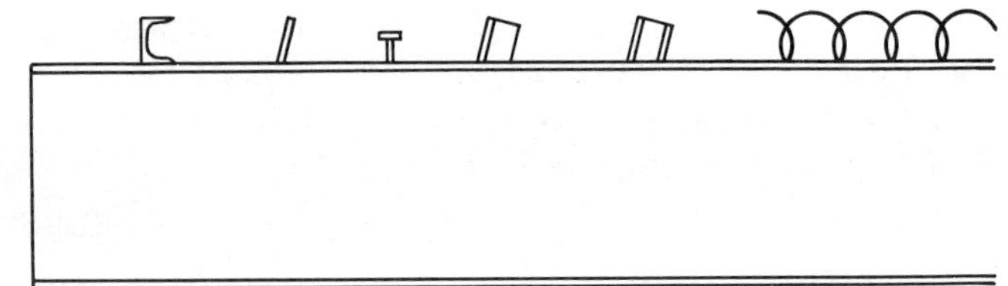

FIGURE 7.1 Several types of shear connectors.

In timber-concrete construction, the shear developer plate has been widely used in slab bridges. In timber-concrete T-beam construction, notching of the top flange of the timber member provides a good shear connection. However, this type of connection should be supplemented by some heavy spikes or a few lag bolts to prevent slab uplift, especially at the beam ends. Figure 7.2 shows a notched beam.

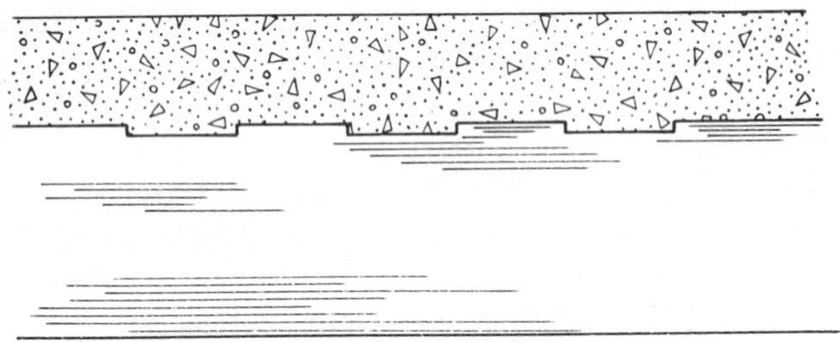

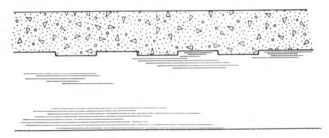

FIGURE 7.2 Notched timber beam.

7.3 How Much Natural Bond?

The early investigators of composite construction clearly showed the existence of a natural bond between steel and concrete. Later tests at Iowa State College included one greased beam. Before testing, the flange was removed, and both the beam and slab were cleaned. As a result of these tests and earlier work, Caughey concluded that the amount of natural bond was largely indeterminate and recommended a mechanical connector. Knowles carries this point much further and concludes that if properly designed shear connectors are provided, then mechanical connectors are unnecessary because the bond carries the entire horizontal load.

The encased beam uses no mechanical connector. The natural bond between the concrete and the completely encased beam is sufficient to develop composite action.

In building work, where loads are primarily static, the natural bond between the beams and slab furnishes a valuable reserve of strength. In bridge work, the moving loads and impact probably destroy the natural bond after a very few load cycles.

7.4 Uplift

Virtually every specification writing body recognizes the tendency of the slab virtually to separate from the beam under some types of loading. During investigations of bonded-aggregate composite beams, the author found that concentrated loads at midspan often caused a visible loss of bond at the beam ends at relatively low load levels.

Although most codes recommend the use of a connector designed to prevent uplift, none specifies any computation or recommended limits for the uplift connectors.

The AISC manual has furnished load tables for the headed stud and the channel connector, both of which provide uplift resistance.

The AITC recommends the use of headed nails or spikes partially driven into the wood members to prevent vertical separation of the wood and concrete.

The ACI permits a closed-loop stirrup, which furnishes good uplift resistance, or another configuration of the extended stirrup to provide adequate embedment length.

7.5 Testing

Although shear connectors can be analyzed and designed by principles of statics and material properties, the results may be overly conservative and depend in large measure on the assumptions made in the analysis and design. Examples of some analyses will be shown in a later section.

The most reliable method of determining the strength of a shear connection is by laboratory test. The tests are usually either the ultimate load, push-out type test, as shown in Figure 7.3, or load-slip tests which can be conducted with the same type of specimen. The allowable load for a given connector can be taken by dividing the ultimate load from the push-out test by a suitable factor of safety or by determining the load required to produce a given amount of slip. Figure 7.4 qualitatively relates the amount of slip to the interaction between the beam and slab. Figure 7.5 shows a typical load-slip curve for a connector. The limiting load defined by the load-slip curve is the load that produces a residual

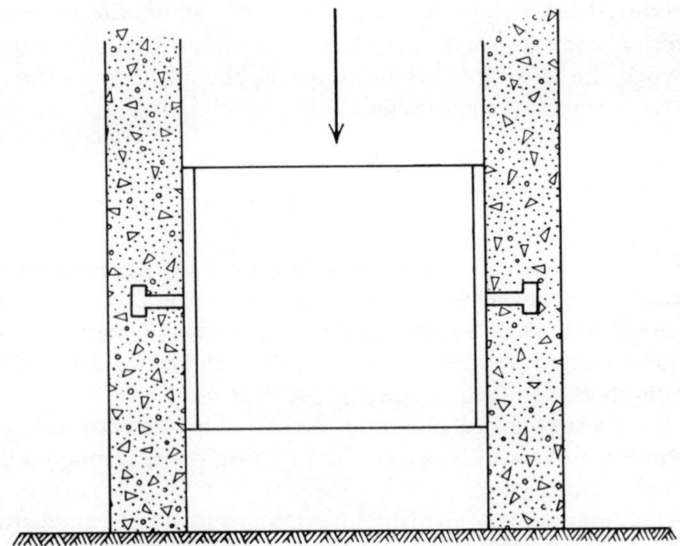

FIGURE 7.3 Push-out specimen.

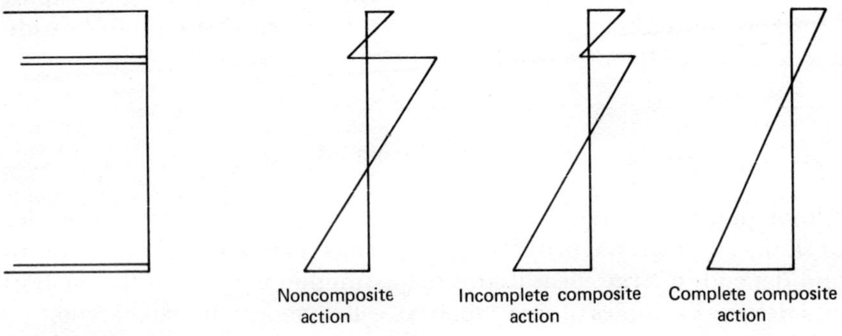

FIGURE 7.4

slip of some given amount, for example, 0.003 in., after removal of the load.

Testing of large-scale composite beam specimens has proved that the ultimate strength of steel-concrete composite beams can be fully developed by a uniform spacing of the connectors along the beam.

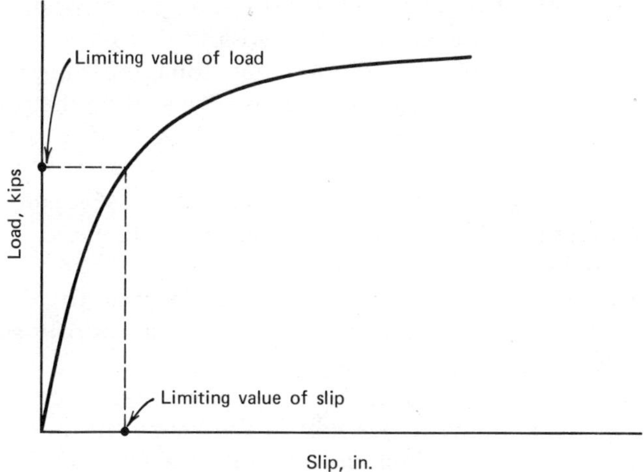

FIGURE 7.5 Load-slip curve.

7.6 Design of the Shear Connection—AISC

The AISC manual tabulates allowable loads for four stud sizes and three channel sizes. Allowable loads for other connectors must be established by a suitable test program.

It is important to note that these allowable tabulated connector loads *cannot* be used with the static equation VQ/I. The allowable loads are based on the ultimate strength of the steel beam and concrete slab, modified by a factor of safety. For full composite action, the total horizontal shear to be resisted between the point of maximum positive moment and joints of zero moment is the lesser of

$$V_h = \frac{0.85 f'_c A_c}{2}$$

$$V_h = \frac{A_s F_y}{2}$$

where f'_c = concrete strength;
A_c = concrete area;
A_s = area of steel beam;
F_y = yield strength of steel.

Note that neither of these formulas takes into account the applied shear force on the beam. Both formulas relate to the ultimate resisting capacity of the beam. One is expressed in terms of steel, and the other is expressed in terms of concrete. This method of connector design states in effect that the beam is at ultimate load and that both the steel and concrete are in a plastic condition. The total horizontal load the connectors must carry is either the compressive load on the concrete or the tensile load in the steel. These loads are then divided by 2 to bring them into the working load range. This type of connector design is a welcome relief for the engineer and contractor because it eliminates the tedious computation of shearing stress at several sections along the beam in order to arrive at a spacing diagram.

Since this design method is based on ultimate strength, a redistribution of stress in the connectors is assumed and a uniform spacing of connectors is used throughout the beam.

It is quite simple to show, using several combinations of beam and slab sizes, that when the neutral axis falls in the beam, the steel value will control the design, and when the neutral axis falls in the slab, the concrete value usually controls the design.

Example 7.1 Full Composite Action—Uniform Loads

Assume the following section properties:

$$A_s = \text{W } 16 \times 40 = 11.8 \text{ in.}^2$$

$$b = 71 \text{ in.; 4-in.-thick slab}$$

$$F_y = 36 \text{ ksi}$$

$$f'_c = 3.0 \text{ ksi}$$

SPAN LENGTH 24 ft

$$V_h = \frac{0.85 f'_c A_c}{2} = \frac{(0.85)(3)(4 \times 71)}{2} = 362.1 \text{ kips}$$

$$V_h = \frac{A_s F_y}{2} = \frac{(11.8)(36)}{2} = 212.4 \text{ kips}$$

Using a $\frac{3}{4} \times 3$-in. headed stud (allowable load per stud = 11.5 kips),

$$N = \frac{V_h}{q} = \frac{212.4}{11.5} = 18.5 \text{ studs on each side of the beam centerline}$$

With one stud at centerline, use a spacing of $(24 \times 12)/(2 \times 18) = 8$ in.

Using a 3-in., 4.1-lb channel, this beam has a flange width of 7 in., so use 5-in.-long sections of channel. Each channel can carry $(4.3)(5) = 21.5$ kips.

$$N = \frac{212.4}{21.5} = 9.87, \text{ say 10 channels on each side of the beam centerline}$$

SPACING $$\frac{12 \times 12}{10} = 14.4 \text{ in.}$$

Space the first connector at 4 in. from the support, and then use a spacing of 14 in.

7.6.1 Partial Composite Action

In cases where it is not possible or feasible to supply enough connectors for complete composite action, a lesser number of connectors may be provided and an effective section modulus computed on the basis of the number of connectors furnished. This situation could occur where one portion of a span was heavily loaded and another section of the same beam carried a much lighter load.

In the preceding sample problem, assume that the center one-half of the span is carrying twice as much load as the outside quarters (Fig. 7.6). For this loading, the moment at the $\frac{1}{4}$ point is 0.714 times the midspan maximum moment. In the outside quarters of the beam, then, the effective transformed section modulus required is

$$(0.714)(S_{tr}) = 0.714(92.8) = 66.25 \text{ in.}^3$$

Now we furnish enough shear connectors to supply the required S_{eff}. According to AISC, the effective section modulus is determined as

$$S_{\text{eff}} = S_s + \frac{V_h'}{V_h}(S_{tr} = S_s)$$

For our sample problem, all the factors in this formula are already determined, except V_h'. V_h' is the horizontal shear capacity we actually supply, that is, the number of connectors supplied times the allowable load per connector. Try nine $\frac{3}{4} \times 3$ studs:

$$V_h' = 9(11.5) = 103.5$$

$$S_{\text{eff}} = 64.6 + \frac{103.5}{212.4}(92.8 - 64.6)$$

$$S_{\text{eff}} = 78.34 \text{ in.}^3 > 66.25 \text{ in.}^3$$

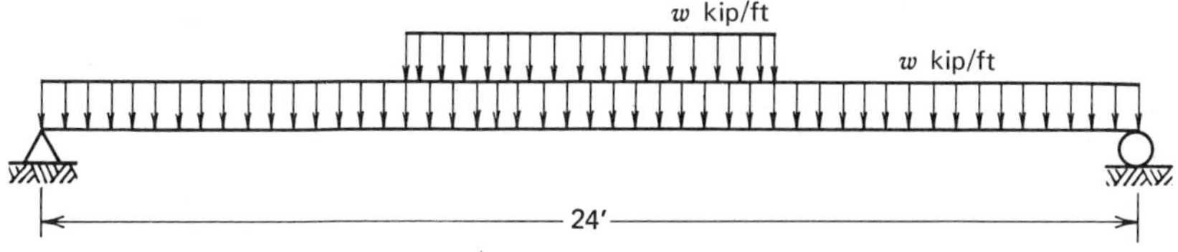

FIGURE 7.6

This is still somewhat overdesigned, and a second trial should be made using a larger spacing.

7.6.2 Negative Bending

In continuous beams that have a negative bending region, the longitudinal reinforcing steel in the slab may be considered to act compositely with the steel beam, provided enough shear connectors are used to tie the beam and slab together. The number of shear connectors required between the maximum negative moment point and the point of zero moment is given by

$$V_h = \frac{A_{sr}F_{yr}}{2}$$

where A_{sr} is the area of reinforcing steel within the effective flange width and F_{yr} is the yield strength of the reinforcing steel.

As an example, a two-span continuous beam is shown in Figure 7.7. The negative moment region extends for 6 ft on either side of the center support, so the formula will give the number of connectors required in a 6-ft length of beam. Using a W 16 × 40 beam with a 4 × 71.5-in. slab, assume No. 8 bars at 6-in. spacing. Use Grade 60 steel for the reinforcing bars.

$$A_{sr} = \text{No. 8 at 6 in.} = 1.57 \text{ in.}^2/\text{ft} \times \frac{71.5}{12} = 9.35 \text{ in.}^2$$

$$F_{yr} = 60 \text{ ksi}$$

$$V_h = \frac{A_{sr}F_{yr}}{2} = \frac{(60)(9.35)}{2} = 280.5 \text{ kips}$$

Using a 3-in., 4.1-lb channel, 5 in. long, the allowable load per connector is 4.3 × 5 = 21.5 kips. The number of connectors required is

$$N = \frac{V_h}{q} = \frac{280.5}{21.5} = 13 \text{ connectors}$$

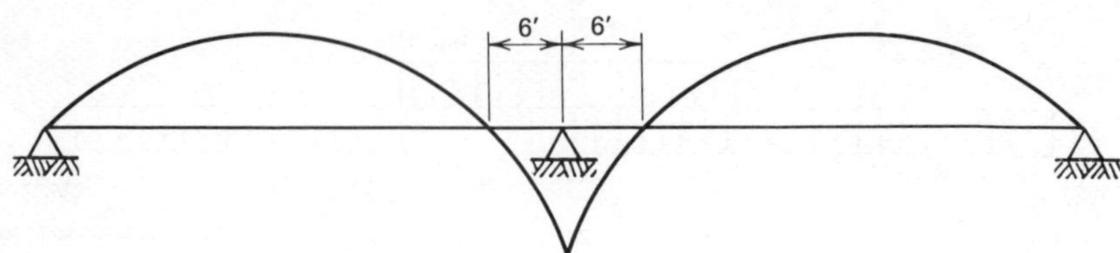

FIGURE 7.7 Two-span continuous beam.

Fitting 13 connectors within the 6-ft interval gives a spacing of $5\frac{1}{2}$ in. At this point welding tolerances must be considered. Figure 7.8 shows the channel at the $5\frac{1}{2}$-in. spacing. For this particular grouping, the welder has enough room between channels to weld both the toe and heel of each channel. If this had been a region of higher negative moment, requiring No. 8 bars at 4-in. spacing, similar calculations would yield a connector spacing of $3\frac{1}{2}$ in. This spacing is too tight to give the welder a decent chance to work. A better choice of connector would be a switch to pairs of $\frac{3}{4} \times 3$-in. studs which would give approximately the same spacing. The studs could easily be attached at this close spacing.

7.6.3 Effect of Concentrated Loads

In the usual building design, floor loads are uniformly distributed and the shear connector spacing is also uniform. However, for members with concentrated loads, the AISC specifications contain an extra requirement. The number of connectors, N_2, required between the concentrated load and the nearest point of zero moment shall not be less than that required by

$$N_2 = \frac{N_1[(M\beta/M\max) - 1]}{\beta - 1}$$

where M = moment (less than the maximum) at a concentrated load point;

N_1 = number of connectors required between the maximum moment point and the point of zero moment, determined by the standard computation;

β = either S_{tr}/S_s or S_{eff}/S_s.

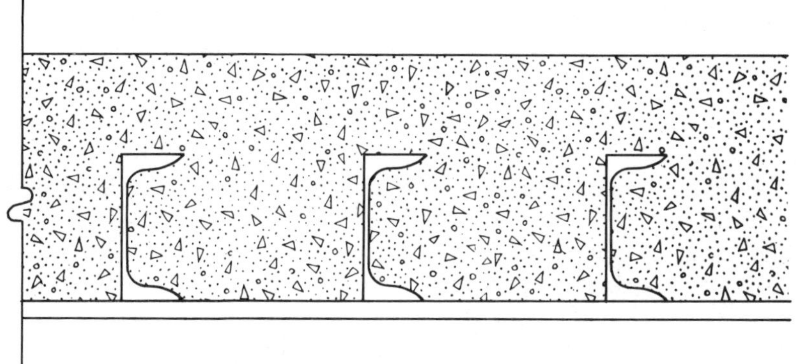

FIGURE 7.8

Note that if the moment under the concentrated load is the maximum moment, the previous formula reduces to $N_2 = N_1$.

As an example for using the concentrated load formula, assume the following:

Uniform load plus concentrated load at the quarter point of a 24-ft span.

$$M = 85 \text{ kip-ft}$$
$$M\max = 96 \text{ kip-ft}$$

Composite beam properties:

$$\text{W } 16 \times 40$$
$$S_{tr} = 92.8 \text{ in.}^3$$

SLAB $\quad 4 \times 71.5$ in.

$$S_s = 64.6 \text{ in.}^3$$

From previous calculations (Example 7.1),

$$N_1 = 18.5 \text{ connectors on each side of the centerline}$$
$$\beta = \frac{S_{tr}}{S_s} = \frac{92.8}{64.6} = 1.44$$

Then
$$N_2 = \frac{N_1[(M\beta/M\max) - 1]}{\beta - 1}$$
$$= \frac{18.5[85(1.44)/96 - 1]}{1.44 - 1} = 11.6 \text{ connectors, say } 12$$

These 12 connectors must be spaced in the 6-ft interval between the concentrated load and the end of the beam. The remaining 6 connectors (of the total 18 required) are to be placed between the concentrated load and the beam centerline.

7.7 Shear Connectors by AASHTO Specifications

The AASHTO specifications require that shear connectors be designed for fatigue and checked for ultimate strength.

The fatigue design follows the static equation VQ/I. The applied shear is usually checked at a number of sections along the beam. The allowable capacity of a shear connector Z_r is determined from the following expressions, which depend on the number of cycles of maximum stress. AASHTO specifies the number of cycles to be used for various span lengths and classes of highways. See the Appendix.

CHANNELS $$Z_r = Bw$$

where

w = length of the channel connector;
B = 4,000 for 100,000 cycles of load
= 3,000 for 500,000 cycles
= 2,400 for 2,000,000 cycles.

STUDS $$Z_r = \alpha d^2$$

where

d = stud diameter, in.;
α = 13,000 for 100,000 cycles of load
= 10,600 for 500,000 cycles
= 7,850 for 2,000,000 cycles

For the welded studs, the ratio of height to diameter (H/d) should be greater than 4.0.

For easy reference, these connector capacities are tabulated below for three sizes of studs and three sizes of channels.

CHANNELS (all sizes) $$Z_r = Bw$$

	Number of Cycles		
	100,000	500,000	2,000,000
Z_r, in kips	4.0 w	3.0 w	2.4 w

STUDS $H/d \geq 4.0$ $Z_r = \alpha d^2$

Z_r, in kips	100,000	500,000	2,000,000
5/8-in. diam.	5.08	4.14	3.07
3/4-in. diam.	7.31	5.96	4.42
7/8-in. diam.	9.95	8.12	6.01

Ultimate Strength Check

After the connectors have been designed on the basis of fatigue, the beam must be checked to determine whether there are enough connectors to develop the ultimate strength of the beam. The number of connectors is checked by

$$N = \frac{P}{0.85 S_u}$$

where N = the number of connectors between the point of maximum moment and the point of zero moment. In continuous beams, use the point of zero dead load moment.

P = either A_sF_y or $0.85f'_cA_c$;

S_u = ultimate strength of one connector.

The ultimate strength of a connector is given by one of the following expressions:

CHANNELS $$S_u = 550\left(h + \frac{t}{2}\right)w\sqrt{f'_c}$$

STUDS $$S_u = 930d^2\sqrt{f'_c}$$

These ultimate strength values are tabulated below for the same three sizes of channels and studs tabulated previously.

CHANNELS (h = average flange thickness; t = web thickness)

C 3×4.1 $$\left(h + \frac{t}{2}\right) = \left(0.273 + \frac{0.17}{2}\right) = 0.358$$

C 4×5.4 $$\left(h + \frac{t}{2}\right) = \left(0.296 + \frac{0.184}{2}\right) = 0.388$$

C 5×6.7 $$\left(h + \frac{t}{2}\right) = \left(0.320 + \frac{0.19}{2}\right) = 0.415$$

	f'_c, in ksi		
S_u, in kips	3.0	3.5	4.0
C 3×4.1	10.78 w	11.65 w	12.45 w
C 4×5.4	11.69 w	12.62 w	13.50 w
C 5×6.7	12.50 w	13.50 w	14.43 w

The ultimate strength for studs is as follows:

	f'_c, in ksi		
S_u, in kips	3.0	3.5	4.0
5/8-in. diam.	19.90	21.49	22.97
3/4-in. diam.	28.65	30.95	33.08
7/8-in. diam.	39.0	42.12	45.03

7.8 Shear Connectors for Timber-Concrete

Three types of shear connection are common in timber-concrete composite construction: the lag bolt, the shear developer plate for slab bridges, and the notched top surface of the wood members. Uplift spikes are usually provided in the notched T beam and in the deck bridge which uses the shear developer plate.

The AITC specifications state that if half the top surface of the beam is notched, only half the area remains effective in shear resistance. Consequently, the allowable shear stress value for the wood is reduced by one-half.

Notches in the top surface of the beam should be at least $\frac{1}{2}$ in. deep and preferably $\frac{3}{4}$ in. deep. When the notches are used, grooves are milled the full width of the beam into each vertical face of every notch (see Fig. 7.9). These grooves help to provide uplift resistance, and because of them, the $\frac{3}{4}$-in. notch is usually better than the $\frac{1}{2}$-in. notch.

Since the wood should be notched before it is pressure treated, the number and size of notches can be left as a practical matter for the laminator to decide. The designer need only specify that one-half of the top surface is to be notched. If the laminator has the equipment to mill a 3-in.-long notch the full width of the beam, then two notches per foot will suffice.

If lag bolts are to be used as shear connectors, they should be turned with a wrench, not hammered, into prebored holes. The hole diameter should be approximately two-thirds of the nominal bolt diameter. Recommended values for various wood species can be found in Table 5.10 of the AICA *Timber Construction Manual.* For use as shear connectors, at least one-half of the bolt length should be turned into the member. The extended portion of the bolt should have at least $1\frac{1}{2}$ in. of concrete cover. The bolt holes should be treated with preservative before installation of the lag bolts.

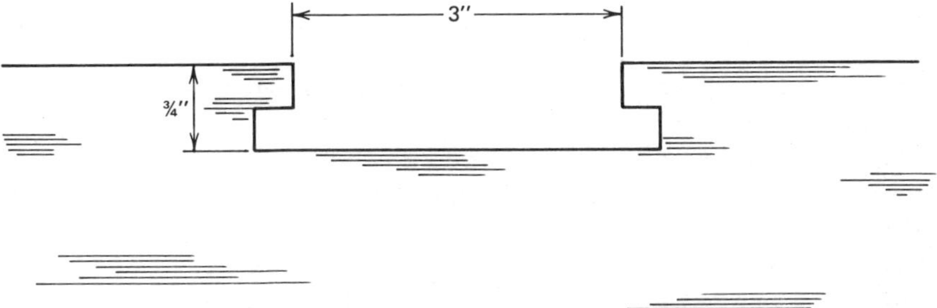

FIGURE 7.9 Milled uplift notched—timber beam.

Example 7.2 Use the same composite T beam used in the example problem of Chapter 6. However, now use this as a building design, carrying 150 psf. Check the shear connection to see whether the notched beam is capable of carrying the shear.

$$F_v = 1/2(145) = 72.5 \text{ psi}$$
$$Q = 2{,}868 \text{ in.}^3$$
$$I = 70{,}783 \text{ in.}^4$$
$$b = 8.75 \text{ in.}$$

LIVE LOAD AT 150 psf × 6 900 lb/ft of beam

LIVE LOAD SHEAR $\dfrac{300 \times 32}{2} = 14{,}400 \text{ lb}$

$$F_v = \frac{VQ}{Ib} = \frac{(14{,}400)(2{,}868)}{(70{,}783)(8.75)} = 67 \text{ psi} < 72.5 \text{ psi} \qquad \text{OK}$$

No additional lag bolts are needed to carry the live load shear. Since this is a building design, there is no need to design temperature connectors if this is an intermediate floor in the building. If this were a roof beam with the slab covered with a roofing material, the temperature differential between the beam and slab could be quite large. Temperature connectors would then be required, as shown in the example in Chapter 6.

Check a shear connection which uses only lag bolts to carry the shear. Referring to the bridge problem in Chapter 6, use the following conditions:

ALLOWABLE LOAD PER BOLT 935 lb

LIVE LOAD SHEAR $V = 18{,}250 \text{ lb}$

$$Q = 2{,}868 \text{ in.}^3 \qquad b = 8.75 \text{ in.} \qquad I = 70{,}783 \text{ in.}^4$$

$$F_v = \frac{(18{,}250)(2{,}868)}{(70{,}783)(8.75)} = 84.5 \text{ psi}$$

NUMBER OF BOLTS $= \dfrac{(84.5)(8.75 \times 32 \times 12)}{935} = 304 \text{ bolts}$

The number of temperature connectors is already computed as 180 bolts in Section 6.1.2.

The total number of bolts required = 180 + 304 = 484 bolts. This number of connectors provides 484/32 = 15+ bolts per foot of span. With this number of bolts per foot, the connectors would have to be placed in pairs or in threes, depending on the edge distance requirements as shown in Table 5.6 of the AITC manual, and as reproduced in the Appendix.

Shear Developer Plates

The shear developer plates are generally used in the timber-concrete slab bridge. The shear developer plates should be driven into precut slots in the upper laminations of the deck. Slots for these connectors should be cut with a special chisel-blade power tool and not with saws or hand tools.

The capacity of these plates has been developed by test. The $\frac{3}{32}$-in. plate shown in Figure 6.8 has a capacity of 1750 lb.

Alternatively, for the deck bridge, the high laminations of the wood deck can be dapped, or notched to provide shear resistance. Check the deck section shown in Figure 7.10 to determine whether dapping the high laminations will provide enough shear resistance. Assume that the applied shear values of 51 and 117 psi have already been computed for midspan and the support, respectively. (Refer to Section 6.2.3 for these calculations.)

Allowable Shearing Stresses

WOOD $\quad$ 90 psi

CONCRETE $\quad 2\sqrt{f'_c} = 110$ psi

The daps will be cut in the high laminations. With the dap length equal to one-half of the dap spacing, the shear areas of the wood and concrete are equal. Using $1\frac{1}{2}$-in. lumber for the deck, there are eight pieces per foot and therefore four high laminations per foot of width of bridge deck.

AT MIDSPAN, SHEAR PER FOOT OF HIGH LAMINATION $\quad \dfrac{51 \times 144}{4} = 1836$ lb/ft

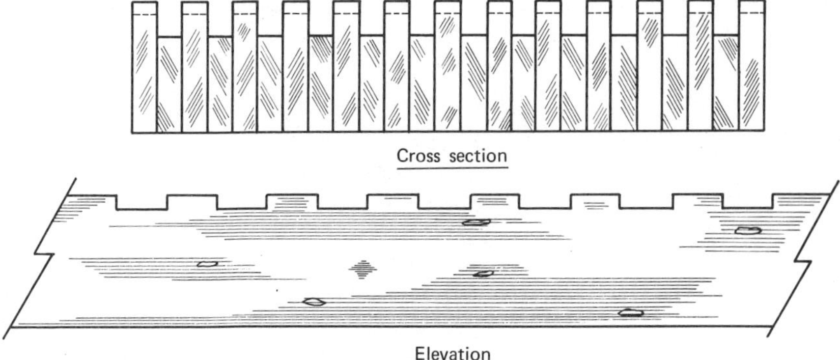

FIGURE 7.10

AT SUPPORT, SHEAR PER FOOT OF HIGH LAMINATION $\dfrac{117 \times 144}{4} = 4212\ \text{lb/ft}$

The allowable shearing stresses of the wood and concrete are very nearly equal, but the wood at 90 psi governs the shear resistance along one high lamination. The area of wood per lineal foot is $6 \times 1.5 = 9\ \text{in.}^2/\text{ft}$.

$$\textit{Resistance} = 9 \times 90 = 810\ \text{lb/ft} < 1836\ \text{lb/ft}$$

so this structure must be provided with shear developer plates.

7.9 Concrete-Concrete Shear Connectors

The shear connection in the concrete-concrete beam comes from two sources:

1. Intentional roughening of the top surface of the precast unit.
2. Ties or extended stirrups across the beam-slab interface.

Intentional roughening means roughening the top surface of the precast member to a full amplitude of $\frac{1}{4}$ in. This intentional roughening is best done in the casting yard before the beams are shipped to the site.

Roughening can be done in several ways. Sand blasting and the needle gun both do a good job. If sand blasting is used, it makes much more sense to do the job in the casting yard under controlled conditions. Another possible method of accomplishing the roughening is to cast the beam units upside down, and coat the bottom of the form (now the top of the beam) with a retarding agent. After the beams are stripped, the top surface of the beam can be hit with a high-pressure water jet. However, this method is tricky and requires good quality control, and it should only be tackled by an experienced casting yard crew.

Power-driven wire brushes on cured concrete are generally unsatisfactory for roughening. They do the job, but often leave behind a black greasy stain which inhibits bond.

The bush hammer does an effective job of roughening. A simple hand rake, which will make $\frac{1}{4}$-inch grooves in the top of the concrete before it takes its initial set, will also suffice.

An AASHTO Class 5, wire brushed or scrubbed finish will also produce satisfactory roughening. This type of finish is produced by scrubbing the surface of green concrete with stiff wire or fiber brushes using a solution of one part muriatic acid to four parts water. When the scrubbing has progressed enough to provide the required roughening, the surface should be flushed with water, to which a small amount of ammonia has been added, to remove all traces of the acid.

The ultimate strength provision of the ACI code states that if the applied shear computed from $v = V_u/\phi bd$ is less than 80 psi, then either intentional roughening or extended stirrups will suffice. If the applied shear exceeds 80 psi, then both roughening and extended stirrups are required. If the applied shear exceeds 350 psi, the beam should be redesigned, or designed according to Section 11.15 of the ACI code, which covers shear friction.

If the alternate design method is used to calculate the shear, the same formula as above is used, but with unity load and ϕ factors. In this case, the allowable shearing stresses are reduced to 55% of the values given above.

If ties are provided to help carry the shear, a closed-loop stirrup is a good solution. This type of reinforcement is properly anchored into the intersecting components, as required by code. The stirrup provides two steel areas crossing the beam-slab interface, and the closed loop at the top provides good uplift resistance.

Example 7.3 Check a concrete-concrete shear connection using both the ultimate strength and the alternate design methods.

DL 200 ppf
LL 500 ppf
SPAN LENGTH 20 ft
$b_w = 6$ in.
$d = 12.5$ in.

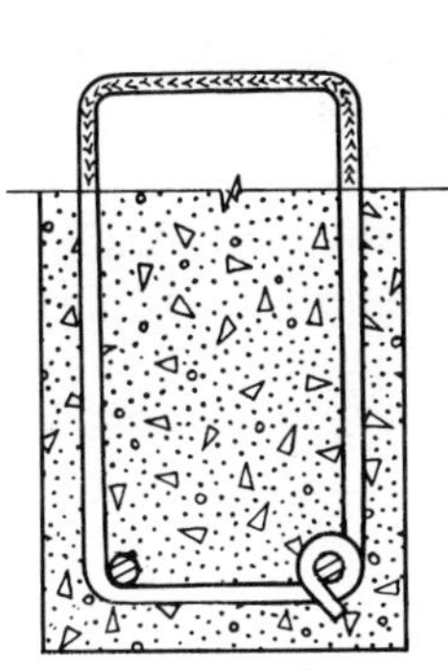

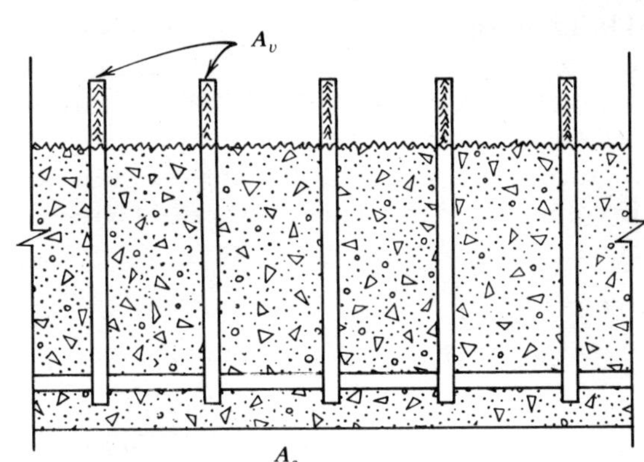

FIGURE 7.11

Ultimate Strength Method

$$V_u = [1.4(200) + 1.7(500)]\frac{20}{2} = 11{,}300 \text{ lb}$$

$$v = \frac{V_u}{\phi bd} = \frac{11{,}300}{(0.85)(6)(12.5)} = 177 \text{ psi} > 80 \text{ psi}$$

Therefore, both roughening and extended stirrups are required.

Alternate Method

$$V = [200 + 500]\frac{20}{2} = 7{,}000 \text{ lb}$$

ALLOWABLE STRESS $= 0.55(80) = 44$ psi

$$v = \frac{V}{\phi bd} = \frac{7{,}000}{(1)(6)(12.5)} = 93 \text{ psi} > 44 \text{ psi}$$

So that both roughening and extended stirrups are required.

7.10 Other Types of Shear Connectors

In steel-concrete construction, the stud and channel are the most widely used types of connectors. For new construction, AISC states that the capacity of any other type of connector must be established by a suitable test program.

However, there are many structures in existence which have used other connectors, such as plates and short sections of wide flange beams. The engineer may be called on to rate some of these existing structures. If the age of the structure is known, it is not too difficult to find the applicable code under which the structure was designed. For example, for a building erected during the 1960s, the design probably followed the requirements of the 1960 Joint Progress Report of AISC and ASCE. Recommended capacities of various connectors according to this report are

STUDS $q = 165d^2\sqrt{f'_c}$

SPIRALS $q =$ allowable load per pitch

$q = 1{,}900d_b\sqrt[4]{f'_c}$

CHANNELS $q = 90(h + 0.5t)w\sqrt{f'_c}$

$h =$ maximum flange thickness

Using 3000-psi concrete,

$\frac{3}{4}$-IN. STUD $$q = 165\left(\frac{3}{4}\right)^2 \sqrt{3{,}000} = 5{,}084 \text{ lb}$$

SPIRAL $\frac{5}{8}$-IN. BAR $$q = 1{,}900\left(\frac{5}{8}\right)\sqrt[4]{3{,}000} = 8{,}788 \text{ lb/pitch}$$

CHANNEL, 5×6.7, 6 IN. LONG

$$q = 90(0.45 + 0.5(0.19))6\sqrt{3{,}000} = 16{,}120\ lb$$

It is interesting to note that these values for the $\frac{3}{4}$-in. stud and the C5 × 6.7 are approximately one-half of the allowable values given for these two connectors in the current AISC specifications.

In the event that no code is available, the engineer may be forced to evaluate a composite structure using only basic principles. The three analyses that follow are patterned somewhat after those shown in the 1953 edition of *Alpha Composite Construction Engineering Handbook.* Assume that the computation of $v = VQ/I$ has been completed and the applied shear, v, is 1.8 kips/in. for all cases.

Spiral

Horizontal force carried by the spiral bar:

$$F = sH = 2A_{sp}f_{sp}$$

Note that in each loop of the pitch, s, there are two bar areas. Using No. 5 bars, $A_{sp} = 0.31$, and using an allowable steel stress of 20 ksi,

PITCH $$s = \frac{2 \times 0.31 \times 20}{1.8} = 6.8 \text{ in.}$$

Tests had shown that the diameter of the spiral loop was relatively unimportant. The dominating factor was the diameter of the bar used to form the spiral.

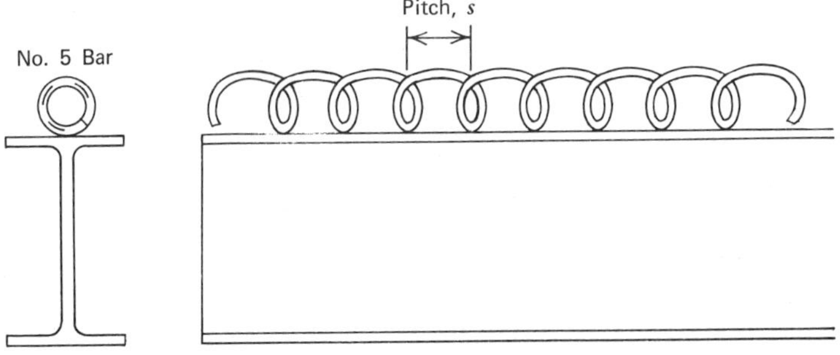

FIGURE 7.12

Inclined Plate Connector

Use a plate 3 in. high, 6 in. long, and $\frac{5}{8}$ in. thick, as shown in Figure 7.13.

Use a basic allowable steel stress of 20 ksi and a concrete stress of 1.35 ksi. These values can be modified depending on the age of the structure. In the 1950s, it was not uncommon to analyze shear connectors using a 50% increase in the basic allowable stresses.

Concrete stresses result from bearing of the concrete against the connector. The assumed concrete stress distribution is shown in Figure 7.14.

CONCRETE, ALLOWABLE LOAD $\quad 1.5(1.35)\left(\frac{1}{2}\times 3\times 6\right) = 18.22 \text{ kips}$

For the steel, allowable load, consider the plate as a short cantilever up from the flange of the steel beam. Determine the resisting moment of

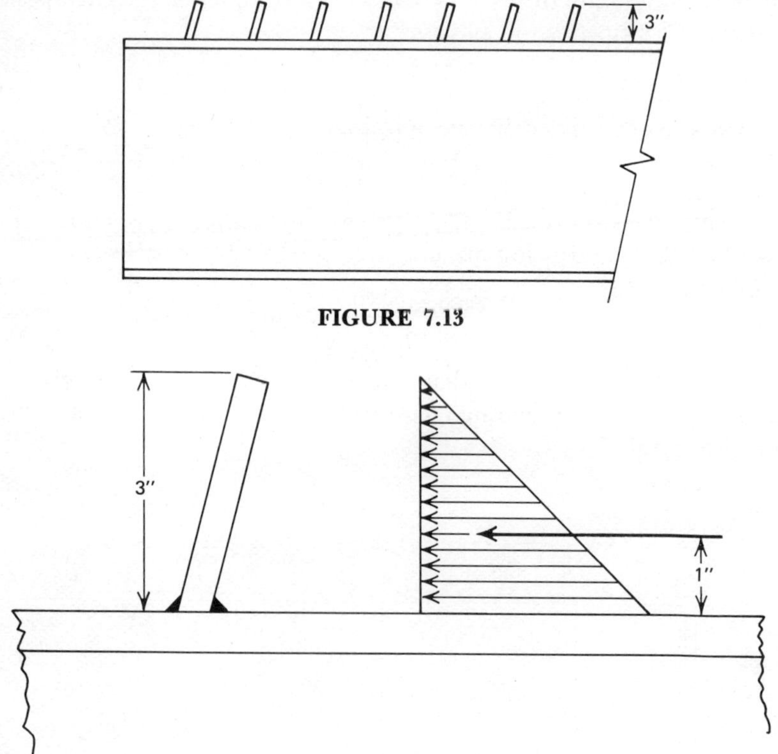

FIGURE 7.13

FIGURE 7.14

the plate and convert it to an allowable load. Shear in the plate itself and the stresses in the welds are seldom critical.

M_r OF PLATE $\quad = (1.5)(20)\left(\frac{6 \times \overline{0.625^2}}{6}\right) = 11.72$ kip-in.

The triangular stress distribution of the concrete on the 3-in.-high plate places the centroid of the load at 1 in. above the top flange of the beam.

$$M_r = \text{load} \times 1 \text{ in.}$$

LOAD $\quad$ 11.72 kips

$$11.72 \text{ kips} < 18.22 \text{ kips} \qquad \text{so the steel governs}$$

PLATE SPACING $\quad \frac{11.72}{1.8} = 6.5$ in.

Channel Connector

The critical point for bending in the channel section is the point where the toe of the lower flange fillet meets the web. In the earlier days of composite construction, research had not yet established the higher allowable values in use for connectors. Consequently, the stockier sections with thicker webs were more frequently used as shear connectors.

Use C3 × 5, 6 in. long. Because the top flange of the channel section provides a good grip on the concrete, increase the allowable stresses by 75%. Stress is again assumed as a triangular distribution.

CONCRETE, ALLOWABLE LOAD $\quad (1.75)(1.35)\left(\frac{1}{2} \times 3 \times 6\right) = 21.26$ kips

For the steel channel, allowable load, determine the resisting moment of the section and convert it to an allowable load. The position of the load has already been determined as 1 in. above the beam flange, as shown in Figure 7.15.

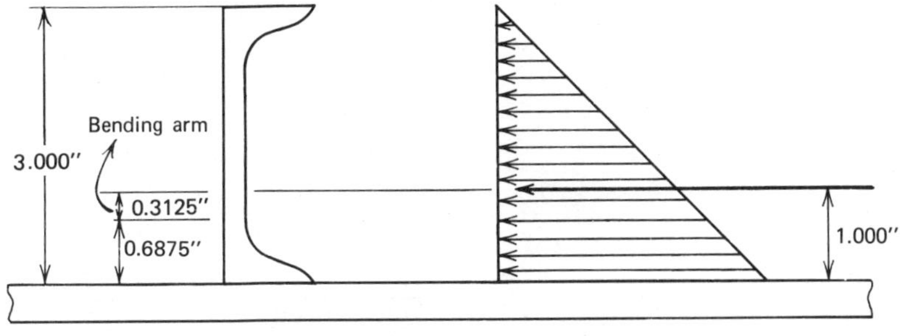

FIGURE 7.15

$$\text{C3} \times 5, Mr = (1.75)(20)\left(\frac{6 \times \overline{0.258}^2}{6}\right) = 2.33 \text{ kip-in.}$$

LOAD × ARM $\qquad M_r$

LOAD $\qquad \dfrac{2.33}{0.3125} = 7.46 \text{ kips} \longleftarrow \text{governs}$

SPACING $\qquad \dfrac{7.46}{1.8} = 4.14 \text{ in.}$

Other shear connectors, such as short stubs of I beams and Z sections, can be checked in a similar fashion. However, the results must largely depend on the validity of the assumptions made in the analysis.

7.11 The Bond Connector

Bond connectors, as the name implies, transmit the horizontal load from steel to concrete by bond, rather than by bearing or bending. No American code currently recognizes the bond connector, as such. It is known, however, that the spiral and the serpentine bar, which formerly were widely used, transmitted the load by bond or by a combination of bond and bearing.

The German Code of Practice recommends for bond connectors an embedment length of 30-bar diameters in the compression zone of the concrete. Of this total length, at least 10-bar diameters must be horizontal. For a No. 8 bar, this embedment length is 30 in. to develop fully the tensile strength of the bar. The bond connectors are considered ineffective in compression. Consequently, they must be properly oriented in order to transmit the horizontal shear.

The ACI code does not recognize the bond connector. However, to develop fully the strength of a reinforcing bar, ACI specifies an embedment length:

$$l_d = \frac{0.04 A_b f_y}{\sqrt{f'_c}}$$

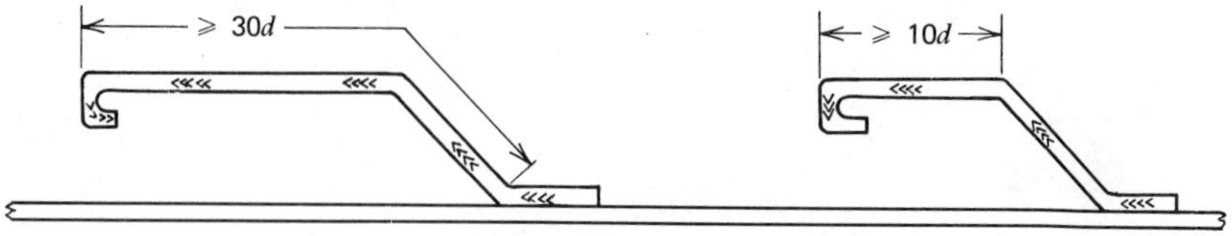

$\geqslant 30d$ $\qquad$ $\geqslant 10d$

FIGURE 7.16

For a No. 8 bar of Grade 60 steel, with 3000-psi concrete, the embedment length is 34.6 in., which is very close to the value from the German code.

Using the alternate design provision of the ACI code,

ALLOWABLE LOAD PER CONNECTOR $\quad 0.79 \times 24 = 18.96$ kips

This same provision of the code specifies doubling the shear values when dealing with the embedment length of reinforcement. Using the previously computed value of 1.8 kip/in.,

SPACING $$\frac{18.96}{2 \times 1.8} = 5.25 \text{ in.}$$

Using a 4-in. slab, this spacing meets the requirements of the German code which state that the spacing should be between $0.7t$ and $2.0t$.

7.12 Lightweight Concrete

For steel-concrete composite beams using lightweight concrete, research has established that modified values of shear connector capacities should be used. A table of coefficients relating these reduced strength values to concrete weight has been shown in Chapter 3.

For concrete-concrete composite beams, no distinction is made in the design of the shear connection, whether normal weight or lightweight concrete is used.

For timber-concrete composite beams, AITC makes no distinction between the two types of concrete. However, research is currently under way which may demonstrate whether reduced connector values should be used for the lightweight concrete.

References

7.1 Ollgard, J. G., Slutter, R. G., and Fisher, J. W., "Shear Strength of Stud Connectors in Light Weight and Normal Weight Concrete," *AISC Journal*, Vol. 8, No. 2, 1971.

7.2 "Table for Selecting Shear Connectors in Composite Design for Buildings," Engineering Aids, Bethlehem Steel Company, 1973.

7.3 "Joint ASCE-AISC Progress Report on Composite Construction," *J. Str. Div. ASCE*, New York, December 1960.

7.4 Viest, I. M., and Siess, C. P., "Design of Channel Shear Connectors for Composite I Beam Bridges," *Public Roads*, Vol. 28, No. 1, 1954.

7.5 Newmark, N. M., Siess, C. P., and Viest, I. M., "Tests and Analyses of Composite Beams with Incomplete Interaction," *Proc. Soc. Exp. Stress Analysis*, Vol. 9, No. 1, 1951.

7.6 Viest, I. M., Siess, C. P., Appleton, J. H., and Newmark, N. M., "Studies of Slab and Beam Highway Bridges: Part IV Full Scale Tests of Channel Shear Connectors and Composite T-Beams," University of Illinois, Eng. Exp. Station Bulletin 405, 1952.

7.7 *Alpha Composite Construction Engineering Handbook*, Porete Manufacturing Company, N. Arlington, N. J., 1953.

7.8 Chinn, J., "Pushout Tests on Lightweight Composite Slabs," *AISC Journal*, Vol. 2, No. 4, 1965.

8

Deflections

The composite beam is stiffer, and consequently it deflects less than the noncomposite beam of the same size. Deflections are computed by elastic analysis using formulas found in any structural handbook. For a given load and span, then, the deflection is inversely proportional to the moment of inertia, and the ratio of the deflection of the noncomposite to composite is the same as the ratio of S_s to S_{tr}. By running a random sample through the AISC composite beam tables, using beams with a full slab and no coverplates, the deflections of the composite beams are roughly from 35 to 55% of the noncomposite beams.

Live load deflection is always carried by the composite section. Dead loads are not so simple. Dead load deflections largely depend on the construction method. If the beam is shored, the dead load is carried by the falsework. If the beam is not shored, the dead load is carried by the bare beam and dead load deflections should be checked.

8.1 What is Dead Load?

We are not actually concerned with a definition of dead load. The really pertinent question is: How much load must be carried by the bare beam? Certainly the weight of the formwork for the slab, wet concrete, and reinforcing bars must be carried. The weight of the bare beam may or may not be included, depending on camber and construction sequence, because the beams may be already in place when the finished grades for the top of the slab are established. In bridge work, loads such as the sidewalk and railing are dead loads in the sense that they are permanent loads. However, these loads are not applied until after the beam has become composite. Construction loads are also carried by the bare beam. Many specifications, especially state specifications for bridges, include a provision for a construction live load of 50 psf on horizontal surfaces. Crowning and superelevation of bridge slabs must also be considered. Generally, the slab thickness shown on the plans is the minimum design thickness. Extra concrete for crown or superelevation may add an extra 10 to 12 psf to be carried by certain beams.

8.2 Monitoring Dead Load Deflections

Dead load deflections should be monitored, whether the beam is shored or not. This can be done with "story poles," "tattletails," or preferably an engineer's level. If shoring is used, the contractor, by specification, bears full responsibility for the falsework. Most falsework systems will show at least some deflection.

Beam elevations should be checked before concrete placement begins. If the shores are forced in too tightly at midspan, giving the beam an upward deflection, the result may be a slab which is level, but less than the design thickness. The contractor, if he is using screw-jack-type shoring, may find it extremely difficult to remove the jacks after the slab dead load is placed. In some cases, it may become necessary to jack the beam upward a slight additional amount in order to remove the falsework. In any event, the contractor wants to remove the falsework as early as possible. It is imperative that the stresses in the falsework be relieved gradually. The beam will deflect under its load, and this load, if rapidly applied, could break the bond between the shear connectors and the uncured concrete.

If shoring is not used, the bare beam will deflect under the load of the wet concrete. If the elevations for a level slab are being shot from a fixed bench mark, this midspan deflection means more concrete must be added to bring the slab up to grade. This additional concrete is more load, which causes more deflection and a slab thicker than the design at the section of maximum positive bending moment.

8.3 Long-Term Creep Deflection

Under long-term loading, the concrete slab in a steel-concrete beam undergoes a complex interaction of creep and stress relaxation. Creep is well known as a continuing deformation under a constant load (Fig. 8.1).

However, the factor which more strongly affects the slab is stress

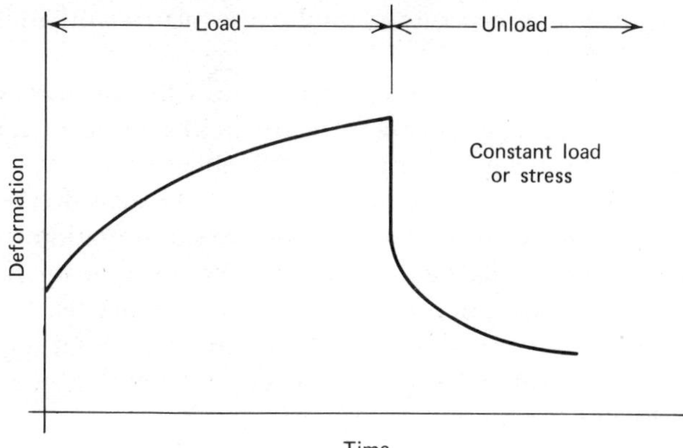

FIGURE 8.1 Creep curve.

relaxation, which is a decay or loss of stress under constant deformation (Fig. 8.2). These two factors interact and actually show up in the composite beam.

In the steel-concrete composite beam, the steel and concrete have creep and stress relaxation properties which are radically different. Practically speaking, the concrete is affected by the long-term loading and the steel is not. The net effect with time is a loss of part of the composite action. Stresses in the slab decrease and the bottom flange stresses in the steel beam increase. Because of the shape of the stress relaxation curve, these effects flatten out after a year or two, with

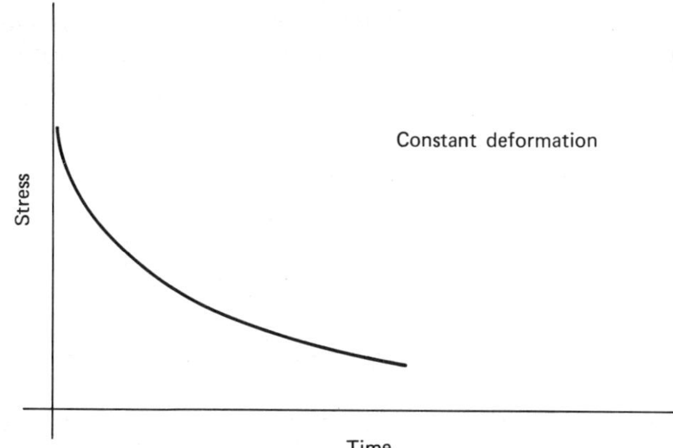

FIGURE 8.2 Stress relaxation curve.

generally only minor effects on the total stress picture and an increase in deflection of 15 to 20%.

In the timber-concrete beam, the wood and concrete have creep and stress relaxation characteristics which are in the same order of magnitude, so that long-term loading has little effect on the stresses in the composite member. However, deflection in wood is also affected by time. Short-term deflections are elastic. Long-term continuation of the load results in an additional plastic deformation. Removal of the load results in recovery of the elastic deformation but not the plastic deformation. Tests have shown that the plastic deformation under long-term loading is one-half of the elastic deformation for glued laminated members and for seasoned solid timbers.

In the concrete-concrete composite member, long-term loading must also be considered. If shoring is used, the composite member may be considered as equivalent to a monolithic cast-in-place member for deflection purposes. The effects of shrinkage and creep can be accounted for by using an amplification factor to compute the additional deflections due to long-term effects. If shoring is not used, ACI states that if the member meets minimum thickness requirements, deflections occurring after the beam becomes composite need not be calculated. However, the precast beam must be investigated for long-term effects prior to the beginning of effective composite action.

8.4 Deflection of Steel-Concrete Beams

The AISC specifications give limiting depth to span ratios which may prevent many deflection problems. However, deflection calculations should still be made. The recommended depth to span ratios are

$$\frac{1}{22} \text{ for } F_y = 36 \text{ ksi}$$

$$\frac{1}{16} \text{ for } F_y = 50 \text{ ksi}$$

The depth used here is the depth from the top of the slab to the bottom of the steel section.

Allowable deflections depend on usage and the applicable building code. One widely used requirement is that deflection be limited to 1/360 times the span length, if a plastered ceiling is to be used below the beam. AASHTO limits deflections due to live load plus impact to 1/800 times the span for steel members.

Deflections can be calculated using familiar formulas of structural mechanics. However, tests of steel-concrete composite beams show deflections about 15 to 20% higher than the computed values. This result may be partly due to shear deflections. Because of the increased stiffness of the composite members, the shear deflections may become appreciable. These will be checked in a later example. In usual practice, shear deflections can be neglected unless actual deflections from the simple moment computations are close to the allowable values.

It becomes more important to check deflections in coverplated sections. Coverplates increase strength, but they are best used when deflections are not critical.

The effects of creep and stress relaxation on long-term deflections are accounted for by using a modified value of the modular ratio in deflection computations. AISC recommends a value of $2n$ and AASHTO specifies a value of $3n$.

Example 8.1 Building Design—AISC Specifications
Allowable Live Load Deflection, L/360

SPAN 32 ft
BEAM SPACING 8 ft
4-in. concrete slab
$f'_c = 3.0$ ksi
A-36 steel
$n = 9$

Loading

LIVE LOAD 120 psf
CEILING AND APPLIED FLOOR FINISH 30 psf

Loads on Steel Section (construction loads)

SLAB	4/12 × 8 × 150 pcf = 400 ppf
BEAM WEIGHT (est)	= 40 ppf
	440 ppf

$$M_D = \frac{0.44 \times \overline{32}^2}{8} = 56.3 \text{ kip-ft}$$

Loads on Composite Section

LIVE LOAD 120 × 8	= 960 ppf
FLOOR AND CEILING 30 × 8	= 240 ppf
	1200 ppf

$$M_L = \frac{1.2 \times \overline{32}^2}{8} = 153.6 \text{ kip-ft}$$

Composite Beam Properties

W 18 × 45 with 4-in. slab

$I_{tr} = 1860 \text{ in.}^4$

$S_{tr} = 112 \text{ in.}^3$

$I_s = 706 \text{ in.}^4$

$S_s = 79.00$

$y_b = 16.65 \text{ in.}$

DEAD LOAD DEFLECTION $\quad \Delta = \frac{5}{384}\frac{wL^4}{EI} = \frac{(5)(0.44)(32)^4(1728)}{(384)(29 \times 10^3)(706)} = 0.51 \text{ in.}$

LIVE LOAD DEFLECTION $\quad = \frac{(5)(1.2)(32)^4(1728)}{(384)(29 \times 10^3)(1860)} = 0.52 \text{ in.}$

ALLOWABLE LIVE LOAD DEFLECTION $\quad = \frac{32 \times 12}{360} = 1.06 \text{ in.}$ OK

Check the same beam using AISC formulas.

DEAD LOAD $\quad \Delta = \frac{M_D L^2}{160 S_s y_{bs}}$

$$= \frac{(56.3)(32)^2}{(160)(79)(8.93)} = 0.51 \text{ in.} \qquad \text{as above}$$

LIVE LOAD $\quad \Delta = \frac{M_L L^2}{160(S_{tr})(y_b)}$

$$= \frac{(153.6)(32)^2}{(160)(112)(16.65)} = 0.52 \text{ in.} \qquad \text{as above}$$

For this particular beam, the long-term creep deflection is probably not significant. However, it will be calculated simply to show the method. AISC recommends using a value of $2n$ to compute the long-term deflections. The AISC also notes that using the values of S_{tr} and y_b from the partial slab tables for beams with a full slab will give a good approximation of the long-term creep deflection.

Check the long-term creep deflection of the W 18 × 45 beam. The effective slab width is 71.5 in. Find the neutral axis using $2n$

	A	×	y	$= Ay$
BEAM	13.2	×	8.93	= 117.88
SLAB	$\frac{4 \times 71.5}{2 \times 9} = \frac{15.89}{29.09}$	×	19.86	= 315.56
				433.45

$$y_b = \frac{433.45}{29.09} = 14.90 \text{ in.}$$

Note that this is close to the partial slab value of $y_b = 14.55$.

I_{tr}

BEAM	$I_o =$	$= 706$
	$Ad^2 = (13.2)(5.97)^2$	$= 470.5$
SLAB	$I_o = \dfrac{71.5 \times (4)^3}{(12)(2 \times 9)}$	$= 21.2$
	$Ad^2 = \left(\dfrac{71.5 \times 4}{2 \times 9}\right)(4.96)^2$	$= 390.9$
		1588.6

This is close to the partial slab value of 1540 in.^3

$$\Delta_L = \frac{(5)(1.2)(32)^4(1728)}{(384)(29 \times 10^3)(1588.6)} = 0.61 \text{ in.}$$

Using the AISC partial slab method,

$$\Delta_L = \frac{(M_L)(L^2)}{(160)(S_{tr})(y_b)}$$

$$\Delta_L = \frac{(153.6)(32)^2}{(160)(106)(14.55)} = 0.64 \text{ in.}$$

which is a difference of only 5%.

Check the shear deflection in the same beam. For deflection due to shear,

$$\Delta = \frac{KwL^2}{8AG}$$

where K = shape constant;
A = transformed area;
G = shear modulus of the steel beam.

TOTAL LOAD ON THE BEAM $w = 1.2 + 0.44 = 1.64$ kpf

TRANSFORMED AREA $A = 13.2 + \dfrac{4 \times 71.5}{9} = 45 \text{ in.}^2$

SHAPE FACTOR K: use 2.5 as the factor for WF beams

SHEAR MODULUS G: since the transformed area is in terms of steel, use G for steel $= 0.4E_s$

$$= \frac{2.5(1.64)(32)^2(12)}{8(45)(0.4 \times 29 \times 10^3)} = 0.012 \text{ in.}$$

This extra shear deflection is not important in our case.

8.4.1 *Deflections in the Encased Beam*

For deflections in the encased beam, calculations are similar to those in the ordinary steel-composite member. The value of moment of inertia used for the positive moment design can be used for deflection calculations.

Designers are now making more frequent use of rigid and semirigid connections, instead of simple beam connections. If the beam-to-column connection is fully rigid, it is not unusual to assume that the positive moment region is a function of the reduced length shown in Figure 8.3. Deflections then should be controlled by the same effective length. In most beams, connections are not fully rigid, even when designed as such. However, in the encased beam, this can be varied by the type of construction. If the beam, the connection, and the column are all fully encased, it can be assumed that the concrete bonds to the steel throughout and the connection then becomes very close to a fully rigid connection.

This factor can be extremely important in deflection computations. Under a uniformly distributed load, deflection varies as the fourth power of span length. In the sample problem of Section 4.4.1, which has a 24-ft span, use of $0.8L$ or 19.2 ft reduces the deflection to half the value for a simply supported beam.

The AISC specifications make no specific mention of deflection of encased beams. However, the specifications do permit deflection computations by any accepted method of structural mechanics.

For general office use, it is probably best to use the familiar simple beam formulas which are conservative.

8.4.2 *Deflections According to AASHTO Specifications*

Under the AASHTO specifications, there are a few changes in deflection calculations as compared to the AISC specifications.

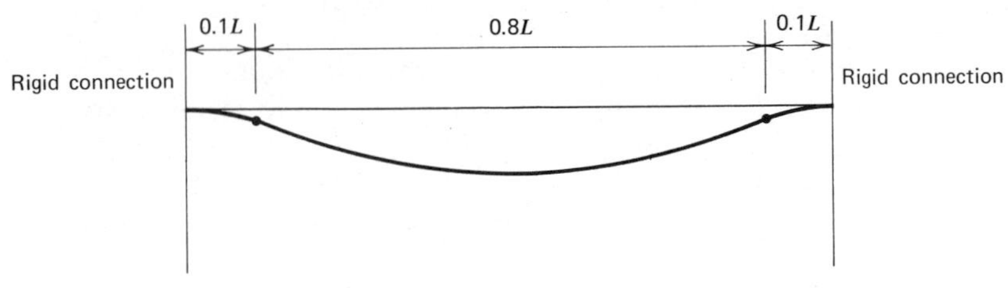

FIGURE 8.3

1. Allowable limit of live load deflection is 1/800 times the span.
2. Dead load on the steel beam and on the composite beam must be defined.
3. Truck load must be correctly positioned for maximum deflection.
4. The use of $3n$ instead of n to account for long-term creep deflections.
5. Live load includes impact.

AASHTO also recommends a depth-to-span ratio for composite beams of 1/25 times the span for A-36 steel.

AASHTO sets no limits on dead load deflection. It is normal practice for the designer to calculate dead load deflections and to furnish the fabricator with a dead load camber diagram. The shrewd designer will check the camber limits for the span length of his particular section and then either provide extra camber or a slight haunch if the member is built up. This is because of an optical illusion. A dead level bridge of reasonably long span, for example, 90 to 100 ft, appears to be sagging when seen from some distance away. By far, the largest number of bridges being built are either on or over expressways, so that they are seen from this perspective. Consequently, the dead load camber diagram becomes very important to the appearance of the bridge.

There are two types of dead load in composite bridge work. Slab dead load and possibly the beam weight, depending on the construction sequence, must be carried by the bare beam. Additional dead load acting on the composite section consists of railing, sidewalk, light standards, an extra wearing course, if used, and so on.

Also in bridge work, crown and superelevation must be checked to obtain the actual slab thickness. Crown in the slab can mean an extra 10 psf acting on certain beams.

Example 8.2 Deflections by AASHTO Specifications

H-S-20 Loading

SPAN 80 ft

A cross section of the bridge is shown in Figure 8.4.

Dead Load on Steel Beam

SLAB $\quad \frac{8}{12} \times \frac{72}{12} \times 0.15 = 0.60$ kpf

STEEL BEAM $\quad$ W $36 \times 260 = \underline{0.26}$ kpf

$\qquad 0.86$ kpf

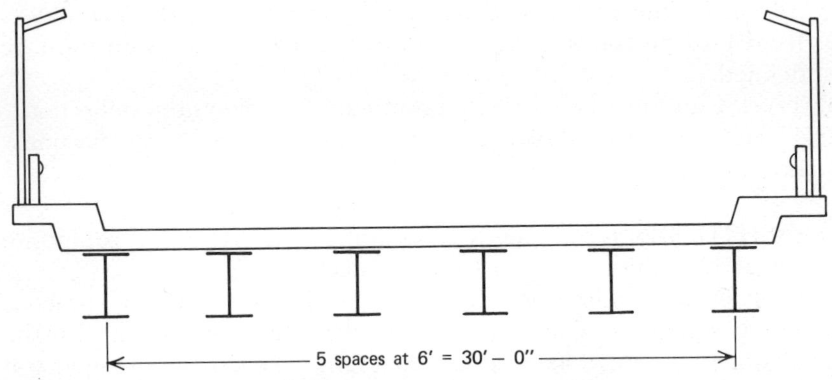

FIGURE 8.4 Bridge cross section.

Dead Load on Composite Beam

SIDEWALK	0.15 kpf
RAILING AND LIGHT STANDARDS	0.04 kpf
FUTURE WEARING COURSE	0.15 kpf
	0.34 kpf

Note that the sidewalk, railing, and light standards are actually mounted almost over the top of the exterior beam. However, AASHTO states that if the longitudinal beam members are adequately tied together by diaphragms, all the beams may be considered together in sharing the load. Consequently, the load for these items was calculated and then distributed over all the beams.

Steel Beam Deflection under Dead Load

$$\Delta = \frac{5}{384}\frac{wL^4}{EI} \qquad I \text{ steel} = 17{,}300 \text{ in.}^4$$

$$= \frac{(5)(0.86)(80)^4 1{,}728}{384(29 \times 10^3)(17{,}300)} = 1.58 \text{ in.}$$

Composite Beam Deflection under Dead Load—Short Term

$$I_{tr} = 33{,}685 \text{ in.}^4$$

$$\Delta = \frac{5(0.34)(80)^4(1{,}728)}{384(29 \times 10^3)(33{,}685)} = 0.32 \text{ in.}$$

Long-Term Creep Deflection of the Composite Beam

Use only that part of the dead load which acts on the composite beam and compute a new value of I_{tr} based on $3n$.

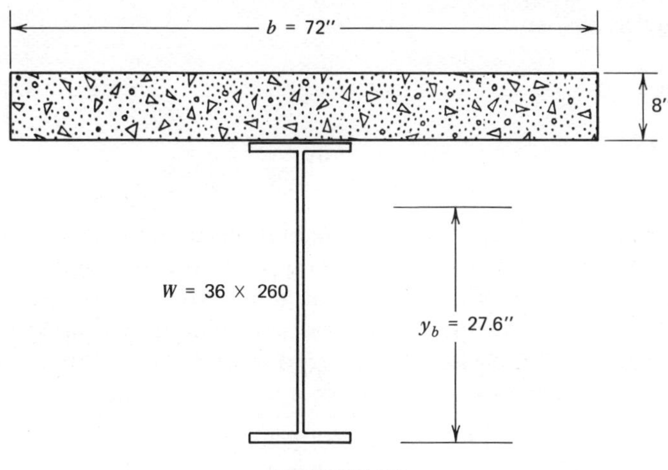

FIGURE 8.5

Section	A	y	Ay
BEAM	76.5	18.12	1868.2
SLAB	$\frac{72\times 8}{3\times 10} = 19.2$	40.24	772.6
	$\Sigma A = 95.7$		$\Sigma A_y = 2640.8$

$$y_b = 27.6 \text{ in.}$$

New value of I_{tr} based on $3n$:

SLAB $\quad I_o - \dfrac{72\times(8)^3}{12\times 3\times 10} = \quad 102$

$\quad Ad^2 - \dfrac{72\times 8}{3\times 10}(12.62)^2 = \quad 3{,}058$

BEAM $\quad I_o - \quad = 17{,}300$

$\quad Ad^2 - 76.5(9.5)^2 = \quad 6{,}904$

$\quad 27{,}364 \text{ in.}^4$

DEFLECTION $\quad \Delta = \dfrac{5(0.34)(80)^4(1{,}728)}{384(29\times 10^3)(27{,}364)} = 0.39 \text{ in.}$

This long-term deflection is roughly 30% higher than the corresponding short-term deflection.

Live Load Deflection

DISTRIBUTION OF WHEEL LOAD PER STRINGER $\dfrac{5}{5.5} = \dfrac{6}{5.5} = 1.09$

IMPACT $\dfrac{50}{L+125} = \dfrac{50}{80+125} = 0.24$

Strictly speaking, for maximum live load deflection, the truck should be placed in the maximum moment loading position as shown in Figure 8.6, but this loading is too cumbersome for general use. However, since live load deflections in composite bridges are seldom critical, it is much simpler to lump all the wheel loads together and place them at midspan. The deflection computation then becomes quite simple.

LIVE LOAD $1.09(4+16+16) = 39.2$ kips

LIVE LOAD PLUS IMPACT $1.24(39.2) = 48.7$ kips

DEFLECTION $\Delta = \dfrac{PL^3}{48EI} = \dfrac{(48.7)(80)^3(1{,}728)}{48(29 \times 10^3)(33{,}685)} = 0.91$ in.

ALLOWABLE $\Delta = \dfrac{80 \times 12}{800} = 1.2 \text{ in.} > 0.91 \text{ in.}$ OK

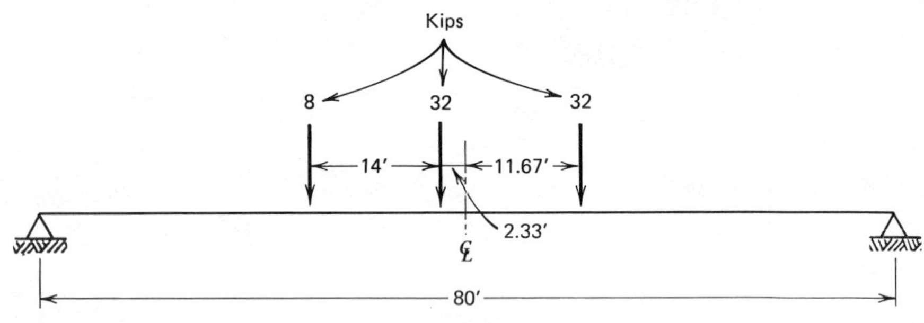

FIGURE 8.6

8.5 Deflection of Concrete-Concrete Beams

Prior to the adoption of the strength method of design by ACI, concrete sections were larger and stiffer and consequently had smaller deflections. With the use of the strength method, plus higher strength steels and concrete, members have become more slender, so that deflection and crack control have become serious considerations.

The familiar formulas for deflection are used with concrete. However, these formulas include E, the modulus of elasticity, and I, the moment of inertia of the member. When designing and building with most other materials, the values of E and I are constant. However, in concrete structures, both of these factors become variables.

The modulus of elasticity varies with load and loading rate. The modulus in tension differs from that in compression. In the early stages of its life, the modulus of the concrete changes rapidly as the concrete gains strength. At early ages, a structure may be strong enough to support the load, but it may deflect enough to cause permanent damage.

With increasing time and application of load, creep and shrinkage change the modulus significantly. Because of all the variables included, the true value of the modulus may actually vary from point to point within the same member under load.

The ACI code recognizes all these variables and specifies $57{,}000\sqrt{f'_c}$ as a safe value of modulus for cured, normal weight concrete.

The moment of inertia of reinforced concrete members has been the subject of a great deal of research. Use of the gross moment of inertia underestimates deflections and use of the cracked section moment of inertia overestimates the deflections. Because of this situation, ACI recommends the use of an "effective moment of inertia" which is between those two limits.

Deflections of concrete members are quite variable. A report of ACI Committee 435 brings out the fact that even under controlled conditions, actual measured deflections can vary as much as plus or minus 20% from the computed values.

The ACI code has greatly simplified the deflection process by furnishing a table of minimum depths of members. Table 9.5(a) from the ACI code is reproduced here as Table 8.1.

If the member meets these minimum depth requirements, deflections need not be computed. If deflections are computed, the allowable limits for deflection may be found in Table 9.5(b) of the ACI code, reproduced here as Table 8.2.

Note that the requirements in both tables are based on construction considerations. The deflections are for members whose deflection will not damage other nonstructural building components, such as partitions and plastered ceilings.

For the long-term creep deflection of the concrete-concrete composite member, an amplification factor, $[2-1.2(A'_s/A_s)]\geq 0.6$, is used to find the additional increment of deflection caused by long-term loading. This additional increment is added to the instantaneous deflection in order to find the total deflection. Note that if no compression steel is used in the beam, A'_s in the amplification factor is zero and the amplification is 2.0.

Table 8.1 Minimum Thickness of Beams or One-Way Slabs Unless Deflections Are Computed[a]

Member	Minimum Thickness, h			
	Simply Supported	One End Continuous	Both Ends Continuous	Cantilever
	Members not supporting or attached to partitions or other construction likely to be damaged by large deflections.			
Solid One-Way Slabs	$l/20$	$l/24$	$l/28$	$l/10$
Beams or ribbed one-way slabs	$l/16$	$l/18.5$	$l/21$	$l/8$

[a] The span length, l, is in in. The values given in this table shall be used directly for nonprestressed, reinforced concrete members made with normal weight concrete ($w = 145$ pcf) and Grade 60 reinforcement. For other conditions, the values shall be modified as follows:
1. For structural lightweight concrete having unit weights in the range 90–120 lb/ft^3, the values in the table shall be multiplied by $(1.65 - 0.005w)$, but not less than 1.09 where w is the unit weight in lb/ft^3.
2. For nonprestressed reinforcement having yield strengths other than 60,000 psi, the values in the table shall be multiplied by $0.4 + F_y/100{,}000$.

The total long-term deflection is then three times the instantaneous deflection. This value of total deflection is somewhat conservative, and the code allows the computation of long-term deflection by a more comprehensive analysis. Several methods are available for these more precise calculations and can be found in textbooks on reinforced concrete design.

Example 8.3 Check the Deflection in Example 4.1

First check the depth of section against Table 8.1. This is unshored construction, so both the precast member and the composite member must be checked.

DEPTH OF PRECAST SECTION 14 in.

DEPTH OF COMPOSITE SECTION $14 + 4 = 18$ in.

MINIMUM CODE DEPTH $l/16 = \dfrac{24 \times 12}{16} = 18$ in.

Table 8.2 Maximum Allowable Computed Deflections

Type of Member	Deflection to be Considered	Deflection Limitation
Flat roofs not supporting or attached to nonstructural elements likely to be damaged by large deflections	Immediate deflection due to the live load, l	$\frac{l}{180}$[a]
Floors not supporting or attached to nonstructural elements likely to be damaged by large deflections	Immediate deflection due to the live load, l	$\frac{l}{360}$
Roof or floor construction supporting or attached to nonstructural elements likely to be damaged by large deflections	That part of the total deflection which occurs after attachment of the nonstructural elements, the sum of the long-time deflection due to all sustained loads, and the immediate deflection due to any additional live load[b]	$\frac{l}{480}$[c]
Roof or floor construction supporting or attached to nonstructural elements not likely to be damaged by large deflections		$\frac{l}{240}$[d]

[a] This limit is not intended to safeguard against ponding. Ponding should be checked by suitable calculations of deflection, including the added deflections due to ponded water, and considering long-time effects of all sustained loads, camber, construction tolerances, and reliability of provisions for drainage.

[b] The long-time deflection shall be determined in accordance with Section 9.5.2.3 or 9.5.4.2, but it may be reduced by the amount of deflection which occurs before attachment of the nonstructural elements. This amount shall be determined on the basis of accepted engineering data relating to the time-deflection characteristics of members similar to those being considered.

[c] This limit may be exceeded if adequate measures are taken to prevent damage to supported or attached elements.

[d] This limit cannot be greater than the tolerance provided for the nonstructural elements. This limit may be exceeded if camber is provided so that the total deflection minus the camber does not exceed the limitation.

Consequently, the deflections of the precast section have to be computed, whereas the deflections of the composite section do not.

SPAN 24 ft

f'_c slab 300 psi

f'_c beam 400 psi

SLAB DIMENSIONS 4×72 in.

STEM SIZE 8×14 in.

$$A_s = 2 \text{ No. } 10 = 2.54 \text{ in.}^2$$

LOADS: PERMANENT DEAD LOAD	0.40 kpf
TEMPORARY CONSTRUCTION LOAD	0.30 kpf
LIVE LOAD	0.90 kpf

First check the precast beam under dead and construction loads.

$$E_c = 57{,}000\sqrt{f'_c} = 3.6 \times 10^3 \text{ ksi}$$

$$y_t = \frac{h}{2} = 7 \text{ in.}$$

$$f_r = 7.5\sqrt{f'_c} = 474 \text{ psi}$$

$$I_g = \frac{8(14)^3}{12} = 1{,}829 \text{ in.}^4$$

$$M_{cr} = \frac{f_r I_g}{y_t} = \frac{474(1{,}829)}{7} = 124{,}000 \text{ in.-lb} = 10.3 \text{ kip-ft.}$$

$$M_a = \frac{wL^2}{8} = \frac{(0.4+0.3)(24)^2}{8} = 50.4 \text{ kip-ft}$$

$$I_{cr} = \frac{8(5.51)^3}{3} + (2.54)(8)(5.98)^2 = 1{,}173 \text{ in.}^4$$

$$I_{\text{eff}} = \left(\frac{M_{cr}}{M_a}\right)^3 I_g + \left[1 - \left(\frac{M_{cr}}{M_a}\right)^3\right] I_{cr}$$

$$= \left(\frac{10.3}{50.4}\right)^3 (1{,}829) + \left[1 - \left(\frac{10.3}{50.4}\right)^3\right] 1{,}173$$

$$= 15.6 + 1{,}163 = 1{,}179 \text{ in.}^4$$

DEFLECTION $$\frac{5(0.3+0.4)(24)^2 1{,}728}{384(3.6 \times 10^3)(1{,}179)} = 1.30 \text{ in.}$$

This is the deflection under dead load plus construction load. The permanent dead load would be $(0.4/0.4 + 0.3)$ times this value, or 0.52 in.

From Table 8.2,

ALLOWABLE SHORT-TERM DEFLECTION

$$l/360 = 24 \times 12/360 = 0.8 \text{ in.} > 0.52 \text{ in.} \qquad \text{OK}$$

The short-term deflections are satisfactory, but now the construction sequence enters the picture. If the precast members are to remain in place for any length of time before the slab is cast, the long-term deflections of the precast member must be investigated. If the precast beams are in place for less than a month before the slab is cast, the long-term effects can probably be neglected. For longer times a check is warranted. For times in the order of months, the ACI amplification factor gives values which are quite conservative.

Check the ACI method.

FACTOR $$= \left[2 - 1.2\left(\frac{A'_s}{A_s}\right)\right]$$

For our case there is no compression steel and thus $A'_s = 0$, so that the factor reduces to 2.0.

LONG-TERM DEFLECTION $(2.0)(0.52) = 1.04$ in.
SHORT-TERM DEFLECTION $= 0.52$ in.
TOTAL Δ $= 1.56$ in.

This figure is probably high. The ACI factor is intended for very long-term sustained loads.

For our beam, the deflections of the composite member need not be checked. If a check were required, the same process would be followed as for the precast beam. The only changes would be to use the T-shaped cross section and the modulus of the 3000-psi slab concrete.

8.6 *Deflection of Timber-Concrete Beams*

In timber-concrete composite T beams, the wood member may be either glulam or solid sawed timber. In the laminated member, camber can be provided as the member is laminated. It is seldom practical to try to camber solid members. It is better practice to limit dead load deflections in the proportioning of the member.

For sustained loads, timber deflections can be handled by a method analogous to the use of $2n$ in steel-concrete. For glulam members, either the sustained loads can be increased by 50% or else two-thirds of the modulus of elasticity can be used in computing the deflection.

For sawed members, a factor is applied to the loads. If the timbers are fully seasoned when erected, the loads are increased by 50%. If the members are not seasoned the loads are increased by 100%.

The allowable deflections in timber members are limited by criteria for either live load only or live load plus dead load, whichever governs. These recommended limits are shown in the table below, taken from the *Timber Construction Manual.*

	Live Load Only	Live Plus Dead Load
Roof Beams		
Industrial	$L/180$	$L/120$
Commercial		
Without plastered ceiling	$L/240$	$L/180$
With plastered ceiling	$L/360$	$L/240$
Floor Beams	$L/360$	$L/240$

In highway bridges, deflection due to live load is usually limited to between $L/200$ and $L/300$, to provide a smooth ride.

Glulam members are normally cambered more than enough to account for the inelastic portion of the deflection curve. In building work, floor and roof beams are cambered to $1\frac{1}{2}$ times the dead load deflection. In bridge work, these members are cambered to twice the dead load deflection for long spans and twice the dead load deflection plus one-half the live load deflection for short spans. The allowable manufacturing tolerance in the cambering of glulam members is $\frac{1}{4}$ in.

Example 8.4 Check the required camber and the deflections in Example 6.1.

This is a glulam member. Calculate the dead load deflection by increasing the dead load 50%.

DEAD LOAD (FROM EXAMPLE 6.1) 505 ppf × 1.5 = 758 ppf

This load is applied to the wood section only.

$$I_w = 19{,}687 \text{ in.}^4 \qquad E_w = 1.6 \times 10^6 \text{ psi}$$

DEFLECTION $$\Delta = \frac{5wL^4}{384EI} = \frac{5(758)(32)^4 1{,}728}{384(17 \times 10^6)19{,}687} = 0.61 \text{ in.}$$

Live Load Deflection

Impact is not included in the computation of stresses for timber members, but impact should be included in deflection computations to insure a smooth riding bridge deck.

For precise calculations, the truck should be placed at the maximum moment loading position with the 16-kip wheel load 1.4 ft from the beam centerline. For simplicity, we can place the 16-kip load at midspan. On this short span, this places the 4-kip wheel load only 2 ft from the beam support where it causes very little deflection.

DISTRIBUTION OF WHEEL LOAD TO STRINGER

$$\frac{S}{5.0} = \frac{6}{5} = 1.2$$

Impact is 30%.

LIVE LOAD $\quad 16(1.2)(1.3) = 25$ kips

DEFLECTION $\Delta \quad \frac{PL^3}{48EI} = \frac{25(32)^3(1{,}728)}{48(1.7 \times 10^3)(73{,}727)} = 0.24$ in.

Camber Computation

Since the span is short, it is probably only necessary to provide the fabricator with a single camber value at midspan. Camber to be provided is "2 times dead load plus 1/2 live load deflection."

TOTAL CAMBER AT MIDSPAN $\quad 2(0.61) + \frac{1}{2}(0.24) = 1.34$ in.

8.7 Deflections in Other Composite Members

Deflections of other types of composite members, such as the timber-concrete deck, the light-gage, metal-deck composite floor, and the flitched beam of wood and steel, are handled by methods similar to the ones illustrated. Composite trusses can be handled by any of the usual truss deflection methods, such as virtual work, which account for the modulus of each member in the computation.

8.8 Effect of Lightweight Concrete on Deflections

The usual effect of lightweight concrete is to increase deflections. In steel-concrete members, a value of $n = 9$ may be used for strength computations, but the actual value of n should be used for deflections. For 3000-psi concrete of various weights, the following values of n may be used:

Concrete Weight	90	95	100	105	110	115	120
Modular Ratio, n	19	17	16	15	14	13	12

Short-term live load deflections using these actual n values can be expected to be about 15 to 20% higher than the deflections using normal weight concrete. This increase should not prove to be any problem in the usual case. However, long-term deflections, which use double this higher n value, should be carefully checked.

For concrete-concrete composite members, Section 9.5.1 of the ACI code makes a special provision for lightweight concrete. Table 9.5(a) gives the minimum thickness of beams of one-way slabs unless deflections are computed. For structural concrete in the 90 to 120-lb range, the tabular values are to be modified as follows:

$$\text{MULTIPLIER FOR TABULAR VALUE} = 1.65 - 0.005w$$

The multiplier shall not be less than 1.09.
For 100-lb concrete, $1.65 - 0.005(100) = 1.15$

For timber-concrete composite beams, no special allowance is usually made for lightweight concrete.

9

Hybrid Design and Construction

The use of high-strength steels in construction is increasing. Steels of different strengths are being combined into hybrid structures. This use of different materials in a single structural member, using each material to its best advantage, is, strictly speaking, composite construction. However, in common usage, the hybrid girder does not become a composite girder until a concrete slab is added.

The AISC manual now lists allowable stresses for steels with yield points of 36, 42, 45, 50, 55, 60, 65, 90, and 100 ksi. Allowable stresses are expressed as a percentage of the yield point. Table 9.1 below, taken from the AISC manual, lists allowable stresses for encased beams, using steels of different strength.

Many possibilities for hybrid structures suggest themselves. Along the span of a beam, different steels can be used in the high and low stress regions. Figure 9.1 shows the dead load bending moment diagram for a beam of three equal spans. Regions are indicated where two different steels might be used.

In the cross section of a built-up girder, the flanges carry most of the bending stress and can be made of higher strength steel (Fig. 9.2). Since the stress distribution across the cross section is virtually linear in the elastic range, the stress in the web at the point where it meets the flange may exceed the yield stress of the web material. However, the flange is intimately attached to the web, and the stronger flange, still in its elastic range, controls the strains in the web. Consequently, AISC concludes that the bending strength of the hybrid member is predictable with the same degree of accuracy as the homogeneous member.

It is possible to change steels along the span as well as across the cross section. One of the earliest successful hybrid girders is a three-span

Table 9.1

	Yield Stress, F_y, ksi								
	36	42	45	50	55	60	65	90	100
Encased beams based on composite section properties, $F_b = 0.66F_y$	24.0	28.0	29.7	33.0	36.3	39.6	42.9	59.4	66.0
Encased beams based on properties of the steel section alone $F_{bz} = 0.76F_y$	27.4	31.9	34.2	38.0	41.8	45.6	49.4	68.4	76.0

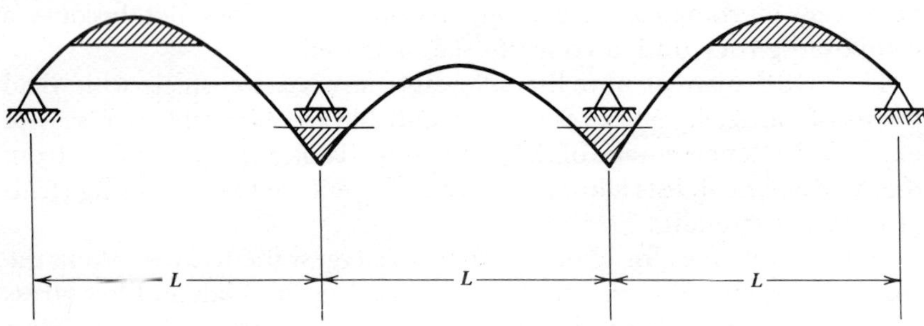

FIGURE 9.1 Three-span hybrid girder. Cross-hatched portions of the spans can use higher strength steel.

girder bridge in Shasta County, California. This girder uses three different steels along the span and varies the web and flange to suit the magnitude of the stresses.

9.1 Modulus of High-Strength Steels

The modulus of elasticity of a material defines the ratio of stress to strain. All the structural steels in use today have virtually the same modulus, as shown in Figure 9.3. Consequently, the high-strength steel,

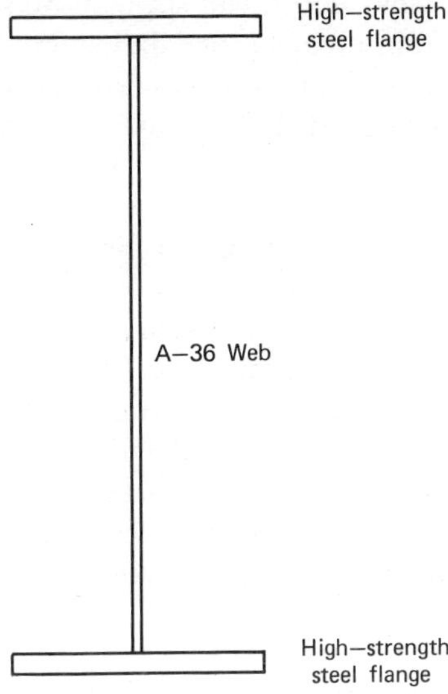

FIGURE 9.2 Hybrid plate girder.

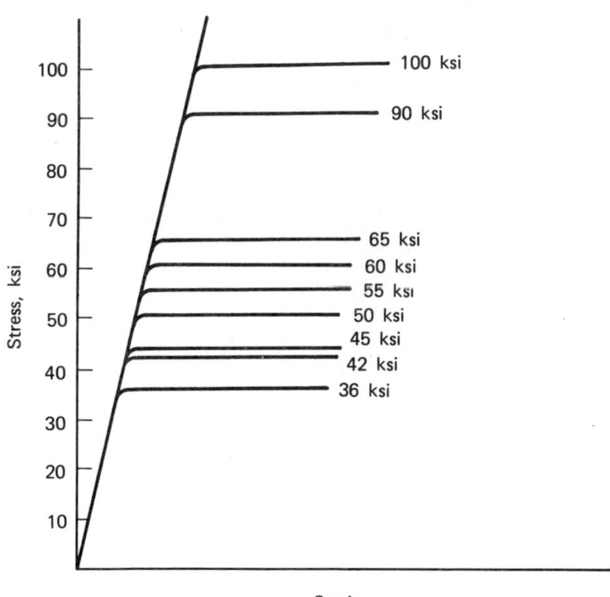

FIGURE 9.3 Modulus curves for different steels.

if stressed to near the yield point, will show more strain than the lower strength steel. More strain means more deflection. In hybrid composite design and construction, then, it becomes especially important to keep a close field check on dead load deflections.

9.2 Code Requirements for Hybrids

The AISC requirements for hybrid design are only minor modifications to the requirements for the design of homogeneous girders. The modifications consider the case of the girder with higher strength steel in the flange than in the web. The modifications use a factor which takes into account the ratio of web yield stress to flange yield stress.

The AASHTO specifications follow the same general line of thinking, but they are somewhat more explicit. Section 1.7.131 (Hybrid Girders) considers two cases:

1. Noncomposite beams and girders that have flanges of the same minimum specified yield strength and a web of a lower minimum specified yield strength.
2. Composite girders that have a tension flange with a higher minimum specified yield strength than the web and a compression flange with a minimum specified yield strength not less than that of the web. This is

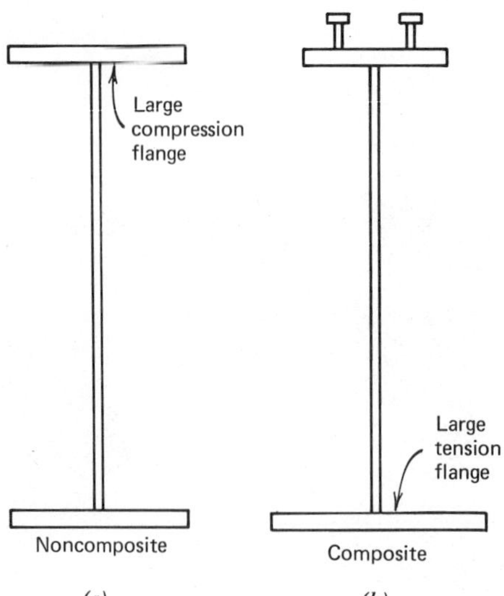

FIGURE 9.4 Girder proportions.

applicable to both simple and continuous girders. The strength of the web shall not be less than 35% of the strength of the tension flange. Figure 9.4 shows the requirements for girder proportioning according to this section of the specifications.

The requirement shown in Figure 9.4*a*, that the compression flange shall be at least as large as the tension flange, relates directly to the lateral buckling of the compression flange, which is especially important under construction loads. The requirement of Figure 9.4*b*, that the compression flange be equal to or less than the tension flange for composite construction, assumes that the compression flange receives adequate lateral bracing until the slab hardens.

9.3 T-on-T Composite Beams

Since the compression flange, according to Figure 9.5*b*, may be less than the tension flange, a logical extension of this thinking is to combine two T sections into a beam. Figure 9.5*a* shows the T-on-T composite beam using a small upper T. In this case, both T's might be of the same grade of steel. Figure 9.5*b* shows the same concept, using two T's of the same size. In this case, the lower T, which is more highly stressed, is made of a higher strength steel, forming a hybrid composite beam. One advantage on the T-on-T design is that there is only one weld, and this weld is in a region of low bending stress, close to the neutral axis. In the hybrid composite, the design is greatly simplified. The T sections are cut from wide flange sections so that two sections of WT 5 × 10.5 placed web to web effectively form a W 10 × 21 section. Section properties can then be determined for the composite beam as if the section were a solid W 10 × 21.

Since all the steels have the same modulus, for noncomposite hybrid design, the modular ratio of one steel to the other equals one, and no transformed section analysis is required.

Example 9.1 T on T Hybrid Composite
Use two T's of the same size, the upper T of 36-ksi steel and the lower T of 50-ksi steel.

SPAN 28′–0″
SPACING 7′–0″
DL = 0.4 kpf
LL = 1.2 kpf

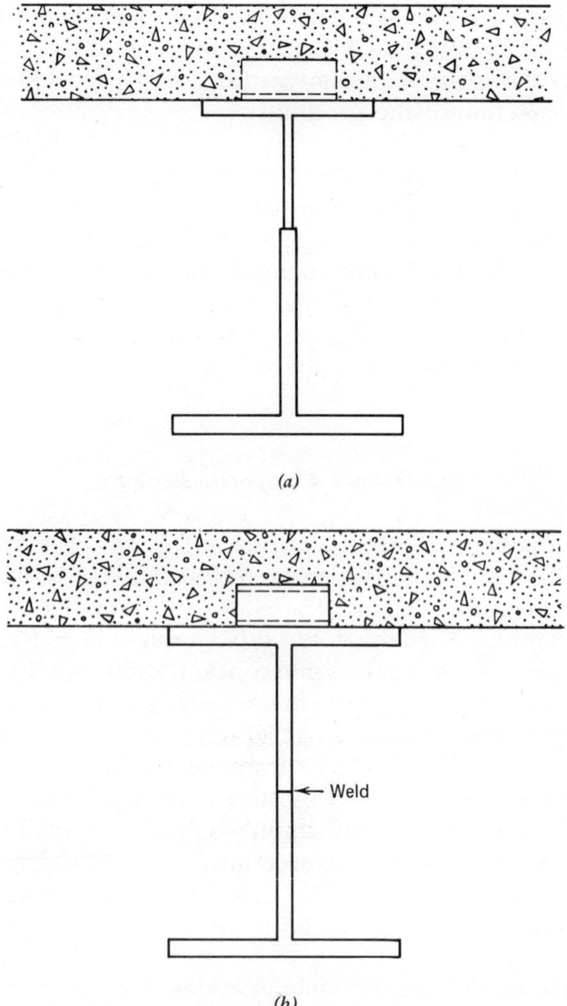

FIGURE 9.5 T-on-T beams.

Since the lower T will be more highly stressed, pick the section from the AISC composite beam tables using 50-ksi steel. There are cautions when using the 50-ksi steel. The concrete stress is more likely to be critical than if A-36 steel were used, especially if the concrete flange is narrow. In our case, with a 28-ft span and a 7-ft beam spacing, the concrete slab flange is well proportioned.

Shored construction usually means higher concrete stresses than unshored, so our beam would probably be of unshored construction. However, the use of higher strength steel usually means smaller steel sections and more deflection, so the dead load deflection will have to be watched carefully.

$$M_D = \frac{(0.4)(28)^2}{8} = 39.2 \text{ kip-ft}$$

$$M_L = \frac{1.2(28)^2}{8} = 117.6 \text{ kip-ft}$$

$$\frac{M_L}{M_D} = \frac{117.6}{39.2} = 3.0$$

$$M_T = 39.2 + 117.6 = 158.6 \text{ kip-ft}$$

Using 50-ksi steel, the allowable bending stress, based on $0.6F_y$, is 30 ksi. The higher allowable stress of $0.67F_y$ for compact sections should not be used because of the longitudinal weld at the center of the section.

REQUIRED $$S_{tr} = \frac{12(156.8)}{30} = 62.7 \text{ in.}^3$$

Using the composite beam tables, a W 14 × 30 section furnishes the following section properties:

$$S_{tr} = 63.4 \text{ in.}^3 \qquad y_b = 13.90 \text{ in.}$$

$$S_s = 41.9 \text{ in.}^3 \qquad b = 70.7 \text{ in.}$$

$$S_t = 223 \text{ in.}^3$$

The section consists of two sections of WT 7 × 15. The upper piece is 36-ksi steel and the lower piece is 50-ksi steel. Three points remain to be checked:

1. Concrete stress.
2. Dead load deflection.
3. The longitudinal weld.

Concrete Stress

$$f_c = \frac{M_r}{nS_t}$$

$$= \frac{156.8(12)}{223(9)} = 0.93 \text{ ksi} < 1.35 \text{ ksi} \qquad \text{OK}$$

Dead Load Deflections

$$\Delta = \left(\frac{5}{384}\right)\left(\frac{wL^4}{EI}\right)$$

$$= \frac{5(0.4)(28)^4(1728)}{384(29 \times 10^3)(290)} = 0.66 \text{ in.}$$

Alternatively, the AISC deflection formula can be used. Two-thirds of an inch is not an excessive dead load deflection.

Longitudinal Weld

For a complete penetration groove weld, the allowable stress in shear is the same as the allowable shear stress in the weaker of our two base metals, the 36-ksi steel. Allowable shearing stress is 14.4 ksi. Use a $\frac{1}{4}$-in. groove weld and E-70 electrodes. Allowable load per inch of weld is 3.6 kip/in. The applied shear on the weld is easily computed from VQ/I. The cross-hatched section in Figure 9.6 illustrates the value of Q.

APPLIED SHEAR $$V = \frac{(0.4 + 1.2)(28)}{2} = 22.4 \text{ kips}$$

$$Q = Ay = 4.42(12.32) = 54.5 \text{ in.}^3$$

$$v = \frac{VQ}{I_{tr}} = \frac{(22.4)(54.5)}{882} = 1.38 \text{ kip/in.}$$

The allowable load per inch of weld is greater than the applied load, so that continuous welding is not required for the purpose of strength. This beam should be shop welded and should be positioned in the shop for flat welding. While continuous welds are not needed for strength,

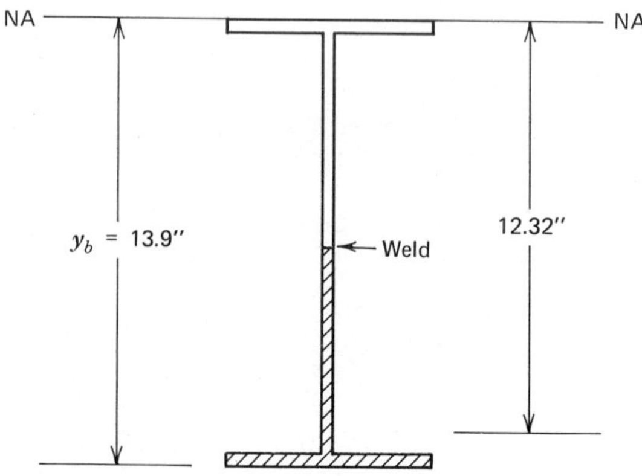

FIGURE 9.6 Shear on center weld.

the fit of the two sections must be good enough so that painting will provide sufficient protection in the sections of the web not being welded.

9.4 Orthotropic Plate Construction

Composite construction is readily used for the strengthening of existing structures. Another method of strengthening existing structures is the use of orthotropic plate construction. Orthotropic plate construction is somewhat analogous to composite construction in that there is a steel deck plate which serves as a large part of the compression flange of the girder, just as the slab does in composite construction.

Orthotropic plate construction is more widely accepted in Europe for new construction than it is in the United States.

The orthotropic plate is structurally efficient, but its design can become quite complex. Usual characteristics of the system are a deck plate and small, closely spaced ribs, or longitudinal stringers. The ribs tie into closely spaced floor beams (7–16-ft spacing), and the floor beams are connected to the main girders. This type of construction is almost always completely welded. There are so many connecting parts that any other type of connection is simply not feasible.

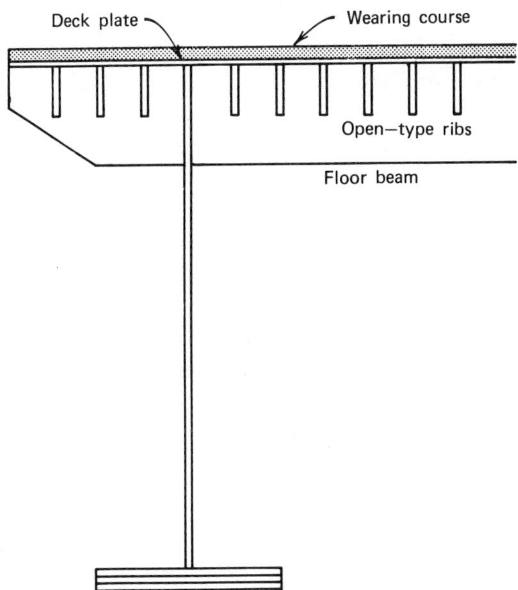

FIGURE 9.7 Open-rib orthotropic plate girder.

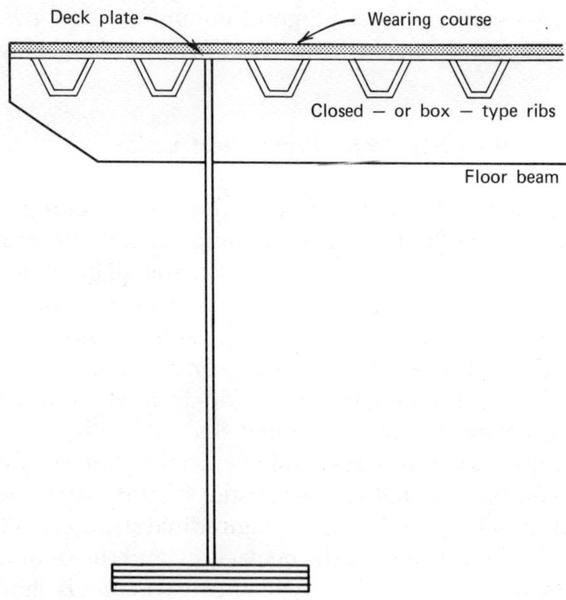

FIGURE 9.8 Closed-rib orthotropic plate girder.

This type of structure can be built with either open or closed ribs. Figure 9.7 shows an open-rib structure and Figure 9.8 shows the closed-rib type.

For reconstruction work with orthotropic construction, the existing bridge slab would be removed. Most standard designs do not have a spacing of floor beams or diaphragms close enough to function as part of the orthotropic system. Consequently, additional floor beams would be added. These can be fabricated as inverted T's. The deck will serve as the top flange. Sections of deck plate can be assembled with the ribs in position and then hoisted into place. This method leaves an overhead weld to be made where the floor beam web attaches to the deck, but this is hard to avoid.

The deck serves in the longitudinal direction as the top flange of the ribs. The deck also serves in the transverse direction as the top flange of the floor beams. The deck and the ribs acting together serve as the top flange of the main girder.

The bridge is usually finished off with a wearing course of asphaltic concrete.

References

9.1 Toprac, A. A., "Strength of Three New Types of Composite Beams," *AISC Engineering Journal*, Vol. 2, No. 1, 1965.

9.2 Column Research Council, "Classification of Steels for Structures," *AISC Engineering Journal*, Vol. 8, No. 3, 1971.

10

The Composite Column

The composite column like its beam counterpart uses two materials to carry the load. Composite columns of steel and concrete are widely known. The concrete-filled pipe column has a long respected history in construction. The steel rolled section encased in concrete is also used. Strictly speaking, every reinforced column is a composite member because it uses both steel and concrete to carry the load. However, in our context only those concrete columns which are reinforced with preformed shapes will be considered.

Some codes formerly used to recognize a separate classification known as a combination column, which was a steel rolled shape encased in concrete. The true composite column had to include longitudinal reinforcing bars as well as the rolled shape.

Under current codes, AISC makes no mention of a composite column. The composite column is covered by the ACI code, which defines the composite column as including all concrete compression members, reinforced longitudinally with steel shapes or tubing and with or without longitudinal reinforcing bars.

Combination of other materials into composite columns is virtually nonexistent. The steel-concrete column is the only one used to any extent. Figure 10.1 shows the usual types of composite columns.

The concrete-filled tube makes an excellent column. An analogy can be drawn between this type of column and the spirally reinforced concrete column. The spiral column is known to have a higher reserve of strength than the tied column. In the tied column at failure load, the exterior concrete spalls off, the longitudinal bars begin to buckle, and the concrete inside the ties begins to fail and shears off, coming out between the ties. In the spiral column, the exterior concrete spalls off,

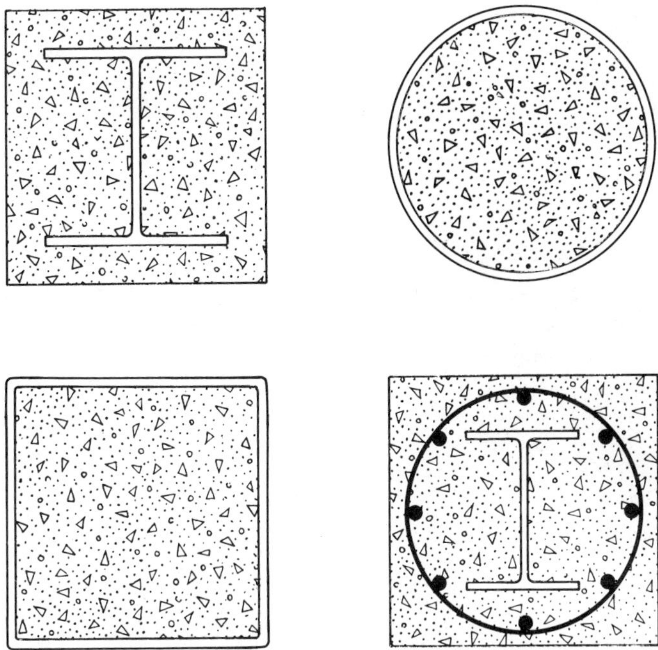

FIGURE 10.1 Types of composite columns.

but the spiral with its close spacing restrains the core concrete, providing a reserve of strength. Figure 10.2 shows typical failures of tied and spiral columns. In the concrete-filled pipe or tube, all the concrete is completely confined, helping to provide strength. In addition, the concrete fill helps to provide stability against buckling in the steel shell that provides the formwork for the casting of the concrete.

The ACI code specifies that any load assigned to the concrete must be transferred to the concrete by members or brackets in direct bearing on the compression member concrete. This direct bearing can be provided by lugs, studs, plates, or sections of reinforcing bar welded onto the steel shape. In effect, we are providing a shear connector to insure that the steel and concrete act together in carrying the load. In the externally encased, structural steel shape, the brackets or lugs are easily welded to the steel member. In the concrete-filled tube, the problem is not so easy. Knowles (3.10) shows that this type of column was used on an interchange bridge in Britain. In this case, the tube was 42-in. O.D., and welded studs were used to transmit the load to the concrete. In the case of a smaller tube, the interior of the tube is not accessible for the entire

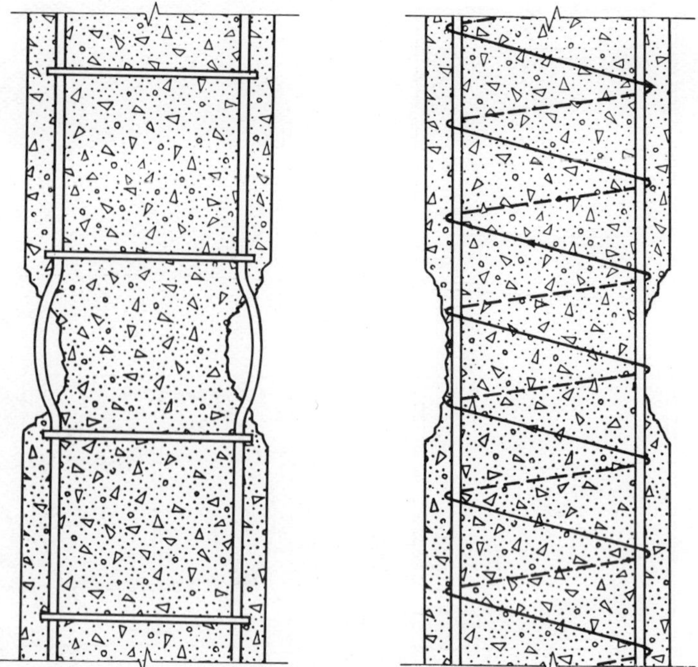

FIGURE 10.2 Typical failures of tied and spiral columns.

column length, so that the lugs or deformed bars would be confined to the ends of the column within the welder's reach.

10.1 Construction Practices

The lugs or brackets which provide the composite action are best welded to the steel section before the column is placed upright.

The procedures for obtaining high-quality concrete in columns are especially important in composite columns because the concrete must be intimately bonded to the bearing units. Ideally, long drops in the placement of concrete should be avoided because concrete tends to segregate. Where it is possible to use them, drop chutes will prevent segregation in many cases. In the tube column, this is probably the best solution. In the encased column, there may not be enough room to place a vertical chute. Windows in the formwork are sometimes used to place concrete in these narrow quarters. Concrete should not be chuted in and allowed to bounce off of the steel section. It is preferable to use a small hopper outside the opening which will allow a better flow of the concrete into the form.

In columns, the concrete should be placed to within 1 ft or so of the top and then allowed to settle for about 1 hr. Concreting must be resumed before cold joints have formed.

Column concrete is usually placed from the top and is normally placed by bucket, rather than by chute. As a rule, vibration is not needed to consolidate different lifts because most one-story columns contain less than 2 yd of concrete, and the concrete bucket may well be larger than this. Vibration is needed in the composite column to insure full contact of the brackets and the surrounding concrete. When placing the last 1 ft of concrete at the top of a column, vibration is necessary to prevent the formation of a cold joint at that point.

10.1.1 Erection Sequence

Consider the case of the encased wide flange section used with a steel-concrete composite beam. If the composite beam is an encased beam, the formwork is relatively simple. The columns, complete with the load-transfer lugs or brackets, are erected and plumbed. In the usual steel frame, column sizes are normally kept constant for a height of two stories, with the column splice about 3 or 4 ft above the floor level, where the ironworkers have easy access to the splice. In the composite column, the steel section can be kept constant for as many as five stories and erected in one piece. The size of the concrete encasement, within limitations, can vary according to design loads. The five-story column, about 60 to 70 ft long, should weigh about 3 tons and can be handled by normal construction equipment. The long column probably does not save any money in erection. The dollars saved in splices will be applied to the extra costs needed to brace and plumb the long column. The savings will be in terms of construction time.

After the columns are set, the steel beams are connected to the columns. The ACI code states that any load not assigned to the concrete shall be developed by direct connection to the steel member.

If the steel composite beam is encased, the column, beam, and slab forms can then be erected, and the casting sequence proceeds as for any reinforced concrete structure.

If the steel composite beam is not encased, the forming of the column around the end of the beam is more difficult. There is usually only about 3 in. of concrete cover around the wide flange section, as shown in Figure 10.3, so that the casting of the concrete around the framing angle

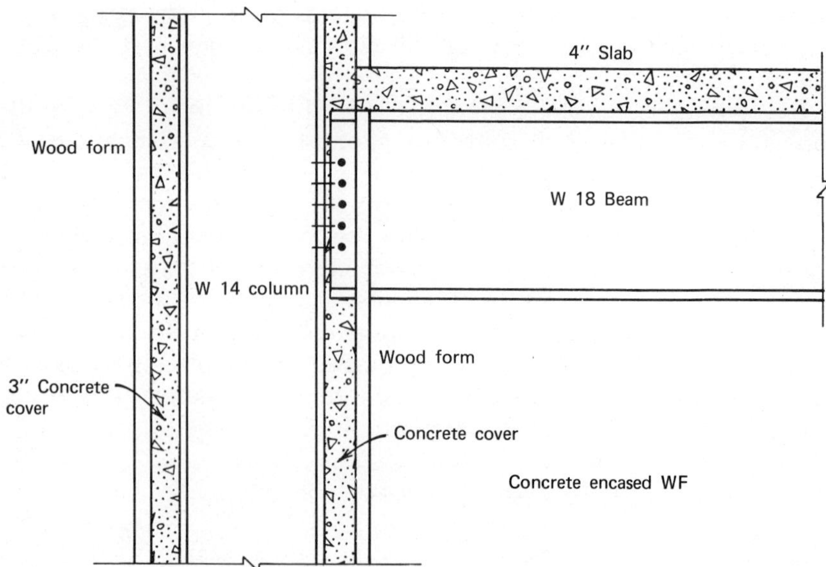

FIGURE 10.3 Beam-to-column connection.

connection and the bottom flange of the beam also requires some special care.

If the structure is of any size, a construction joint will have to be planned into the structure at this point. It is not uncommon in reinforced concrete structures to make this joint at the top of the column where the beam member frames in. In the detail shown, it would be better to cast the columns and slab all at once, if possible. In this way, the concrete can be worked in around the beam end. It is true that this puts the column splice at the point where the beam applies a bending moment to the column, but the continuous steel member precludes the need for any special column splice at this point.

With the concrete-filled steel tube, the problems are slightly different. The tubes are smaller sections and are generally used in shorter structures. The tubes should be prepared with the load-transfer brackets and then erected and plumbed. At intermediate story levels, the beams can rest on top of the columns as shown in Figure 10.4. A two- or three-story tube column is light enough to erect and plumb easily with standard equipment, but filling the column presents a few problems. The concrete must be placed by bucket, preferably with a drop chute, to avoid segregation. However, underwater contractors have successfully tremied concrete down to depths of over 150 ft, so a 30-ft column should

FIGURE 104. Beams across the tops of columns. (Courtesy of Dept. of Physical Plant, Univ. of Cincinnati.

not be too difficult. If the roof beams are to frame over the top of the columns, the columns should be filled to within 1 ft or so of the top and allowed to settle for about 1 hr. When the concreting is resumed, the tube should be overfilled 1 in. or so and then the excess concrete cut off after it has partially stiffened. This will help to insure bearing of the beam members on both the tube and the concrete.

10.2 Code Requirements for Composite Columns

Columns used to be designed, for the most part, as axially loaded members. However, regardless of the type of construction or loading, the usual compression member has at least some bending moment. Recognizing this fact, the ACI code specifies that every compression member be designed for at least a minimum eccentricity. The eccentricity is defined as the bending moment divided by the axial load. Then in a column designed for 200 kips of axial load and 600 kip-in. of bending moment, the eccentricity is

$$e = \frac{M}{P} = \frac{600}{200} = 3 \text{ in.}$$

Figure 10.5 shows the axial load and bending moment replaced by an axial load at the eccentric distance, e. The ACI code specifies that every column shall be designed for at least a minimum eccentricity. This value is $0.05h$ with a minimum of 1 in. for steel-encased composite members.

The overload factors for the design of composite columns are the same as those for beams.

$$U = 1.4D + 1.7L$$

The capacity reduction factor ϕ is lower for columns than for beams. Also, different factors are used for tied and spiral columns. Actual values are not spelled out in the code for composite columns. However, for tube and pipe columns, and for the encased wide flange section, $\phi = 0.75$ and $\phi = 0.70$, respectively, are reasonable values.

The code also gives a formula for the radius of gyration, r, to be used in slenderness calculations:

$$r = \sqrt{\frac{1/5E_cI_g + E_sI_t}{1/5E_cA_g + E_sA_t}} \qquad (10.1)$$

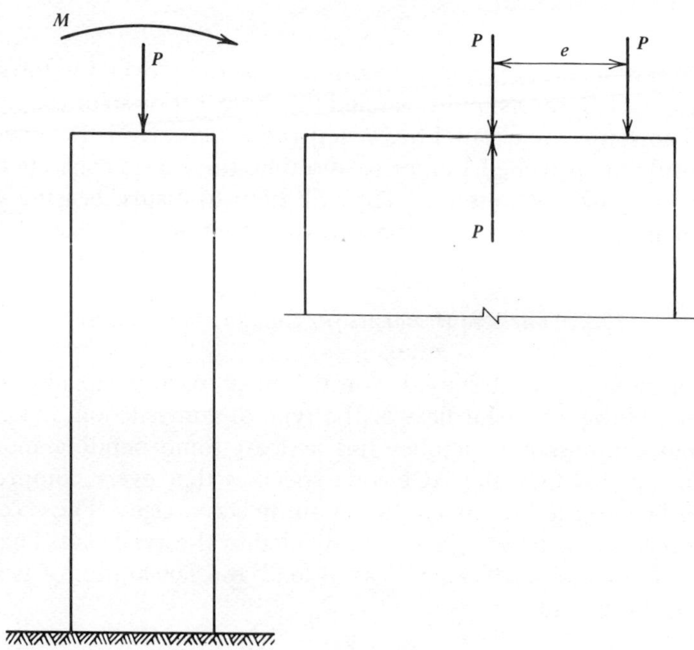

FIGURE 10.5 Eccentric load on column.

The expression above looks complicated, but it is little more than an extension of the expression from strength of materials:

$$r = \sqrt{\frac{I}{A}}$$

The net effect of the formula is a liberalization of Section 10.11.2 of the code, which states that r may be taken as 0.3 times the overall cross-sectional dimension of the column. For a concrete-filled 10×10-in. structural tube,

$$r = 0.3 \times 10 = 3 \text{ in.}$$

Using Formula 10.1,

$$r = 3.75 \text{ in.}$$

Then for a 12-ft-long column, the slenderness ratio of the composite column is 38, as opposed to 48 for the normal 10-in. column.

The code also specifies a formula for EI, the stiffness of the column. This stiffness is used to determine the code's critical load.

10.3 The Interaction Diagram

The axially loaded column with no bending moment and the beam loaded in bending with no axial load are simply the two end points of the general loading case of axial load plus bending. The interaction diagram, Figure 10.6, shows this relationship graphically.

The curve is plotted on axes of axial load versus bending moment. The top of the curve is chopped off because of the code requirement about minimum eccentricity. A number of excellent research reports have been written about interaction diagrams in reinforced concrete columns, but few reports are available on composite columns. The commentary to the ACI code states that the interaction diagram for the concrete-filled steel tube would have the same basic shape as the standard interaction diagram. The diagram is useful because it graphically shows the capacity of the column, and consequently, any combination of axial load and bending moment that falls below the curve is satisfactory for the design.

Theoretically, only three points are needed to plot the diagram:

1. P_0: axial load with no bending.
2. M_0: bending moment with no axial load.
3. The balance point, where the column changes from "tension controls" to "compression controls."

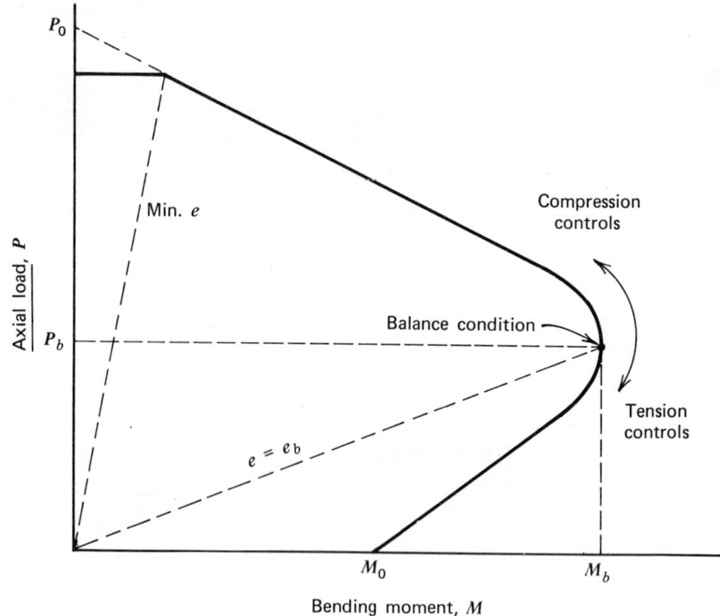

FIGURE 10.6 Column interaction diagram.

This assumes that all the steel has yielded and the concrete has reached its limiting strain, 0.003 in./in., and is about to crush. This situation can exist in the reinforced concrete column with steel in two faces only. In the composite column, the steel reinforcement extends from the compression region into the tension region, so that all the steel is not at yield. The diagram does not make a sharp transition at the balance point, but it takes on the characteristic rounded shape.

Practically speaking, most columns fall in the range of the diagram where compression controls, so it is usually sufficient to plot P_0 and the balance point. A straight line connecting these points is sufficiently accurate to define the compression region. Exact solutions for the complete interaction diagram can be found in standard textbooks on reinforced concrete design.

10.4 The Concrete-Encased Wide Flange Section

Proportion a square column to carry the following:

DL AXIAL 360 kips
LL AXIAL 150 kips
DL MOMENT 80 kip-ft

LL MOMENT 40 kip-ft
CONCRETE f'_c 3,000 psi
STEEL F_y 36,000 psi

Determine the Full Ultimate Load

$$P'_u = \frac{1.4(360) + 1.7(150)}{0.7} = 1{,}084 \text{ kips}$$

$$M'_u = \frac{1.4(80) + 1.7(40)}{0.7} = 257 \text{ kip-ft}$$

Eccentricity of Applied Loads

$$e = \frac{M}{P} = \frac{257(12)}{1{,}084} = 2.85 \text{ in.}$$

The eccentricity is small, so our loads fall within the compression range of the interaction diagram.

The Balanced Condition

This is the loading condition and position of the neutral axis which simultaneously causes a strain of 0.003 in the concrete and first yield in the tensile steel.

$$\frac{x}{d} = \frac{0.003}{F_y/E_s + 0.003}$$

$$x = 0.708d$$

Consider the steel section in four pieces, two flanges and two half-webs.

Selecting the Section

From Figure 10.6,

$$P_b = C_c + C_s - T \simeq C_c$$

$$P_b = (0.85f'_c)(0.85)(0.708d) = 1.54bd$$

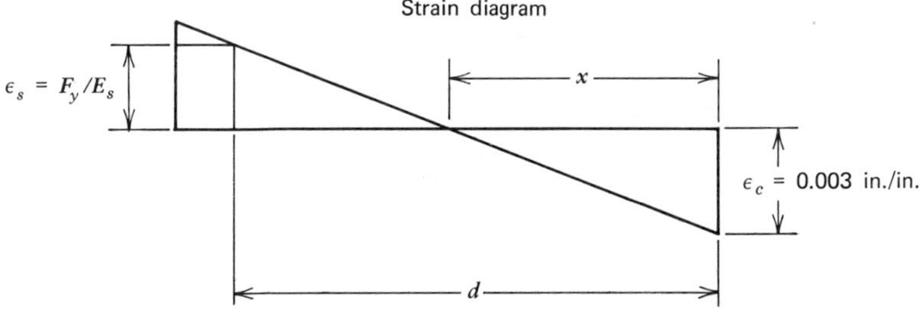

FIGURE 10.7 Column strains.

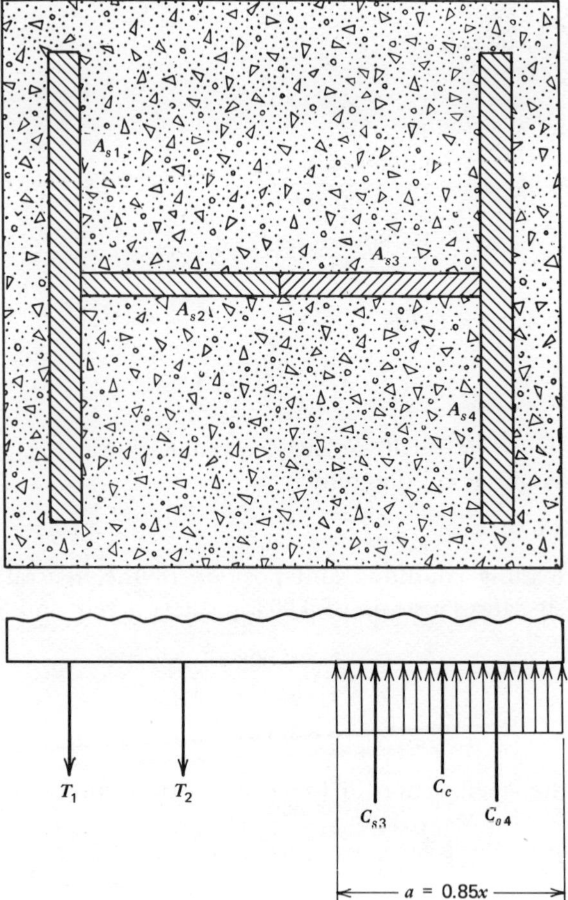

FIGURE 10.8

Find (bd) based on P_0, and then use the approximate formula below to determine the overall column size.

$$bd = \frac{1084}{1.54} = 704 \text{ in.}^2$$

$$(bd)(0.85f'_c) = A_g(0.85f'_c) + (pA_gF_y)$$

$$(704)(0.85)(3.0) = A_g(0.85 \times 3.0) + (0.10 \times 36 \times A_g)$$

Earlier editions of the code specified that the steel column should not exceed 20% of A_g. 10% is assumed above. Solving gives $A_g = 292$ in.² Try an 18×18-in. column with a W 12×106 section.

$$A_g = 18 \times 18 = 324 \text{ in.}^2$$

$$A_s = 31.2 \text{ in.}^2$$

Plot the compression region of the interaction diagram and show the applied loads on the diagram.

A OF CONCRETE $= 324 - 31.2 = 292.8$ in.²

$$P_0 = 0.85 f'_c A_c + A_s F_y$$
$$= (0.85)(3.0)(292.8) + (31.2)(36) = 1870 \text{ kips}$$

Balanced Load (from Fig. 10.7)

$$d = 14.95 \text{ in.}$$

$$x = \frac{0.708}{d} = 10.6 \text{ in.}$$

$$a = \frac{0.85}{x} = 9.0 \text{ in.}$$

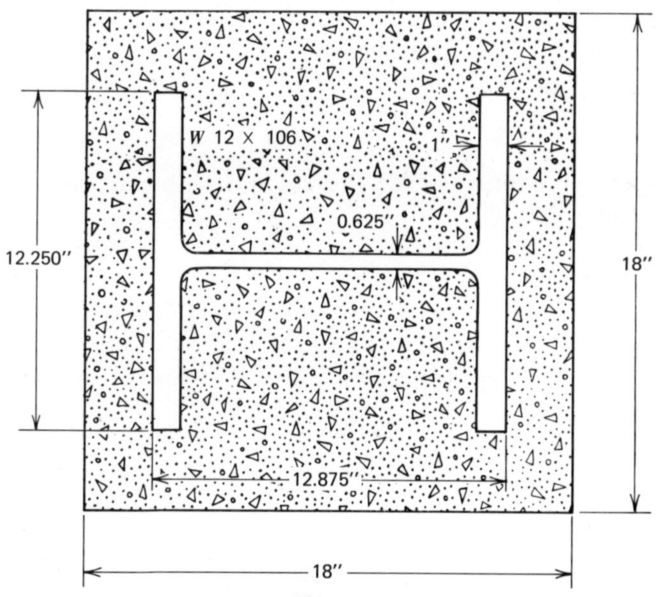

FIGURE 10.9

Now sum the forces and take moments about the plastic centroid of the section. For symmetrical sections, the plastic centroid is the centerline of the section.

Steel Areas

EACH FLANGE $12.23 \times 0.986 = 12.06 \text{ in.}^2$

EACH HALF WEB $\frac{1}{2}[31.2 - 2(12.06)] = 3.54 \text{ in.}^2$

Internal Forces

CONCRETE $C_c = 0.85(3.0)(9.0)(18) = 413.1 \text{ kips}$
HALF-WEB $C_{s3} = 3.54(36) - 3.54(0.85)(3.0) = 118.4 \text{ kips}$
FLANGE $C_{s4} = 12.06(36) - 12.06(0.85)(3.0) = 403.4 \text{ kips}$
FLANGE $T_1 = 36(12.06) = 434.2 \text{ kips}$
HALF-WEB $T_2 = 3.54(36) = 127.4 \text{ kips}$

$$P_b = C_c + C_{s3} + C_{s4} - T_1 - T_2$$
$$= 413.1 + 118.4 + 403.4 - 434.2 - 127.4$$
$$P_0 = 373.3 \text{ kips}$$

Now taking moments about the centerline,

$$Pe = M = (413.1)(4.5) + (118.4)(2.72) + (403.4)(5.94) + (434.2)(5.94) + 127.4(2.72)$$
$$= \frac{7514.7 \text{ kip-in}}{12} = 626 \text{ kip-ft}$$
$$e = \frac{M}{P} = \frac{7514.7}{373.3} = 20.1 \text{ in.}$$

The interaction diagram for the column with our design loads is shown in Figure 10.10.

10.5 The Concrete-Filled Steel Tube

Plot the interaction diagram for a $10 \times 10 \times \frac{3}{8}$-in. structural tube, filled with concrete.

$A_s = 13.8 \text{ in.}^2$ $\quad A_c = 85.5 \text{ in.}^2$

$F_y = 36 \text{ ksi}$ $\quad f'_c = 3.0 \text{ ksi}$

The ACI code specifies that the tube shall have at least a minimum thickness greater than

$$b\sqrt{\frac{F_y}{3E_s}}$$

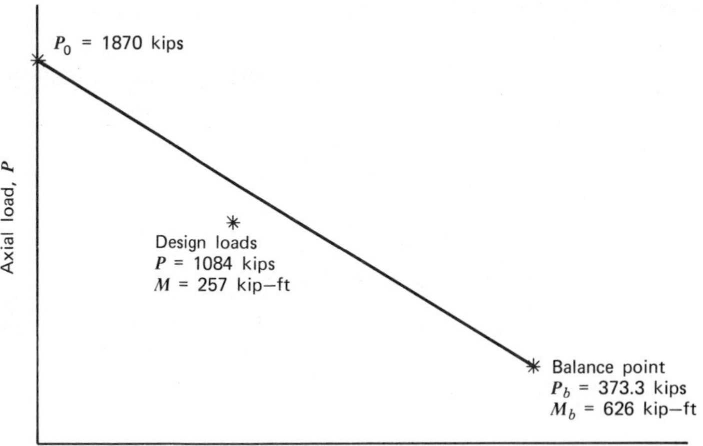

FIGURE 10.10 Interaction diagram—W 12 × 106.

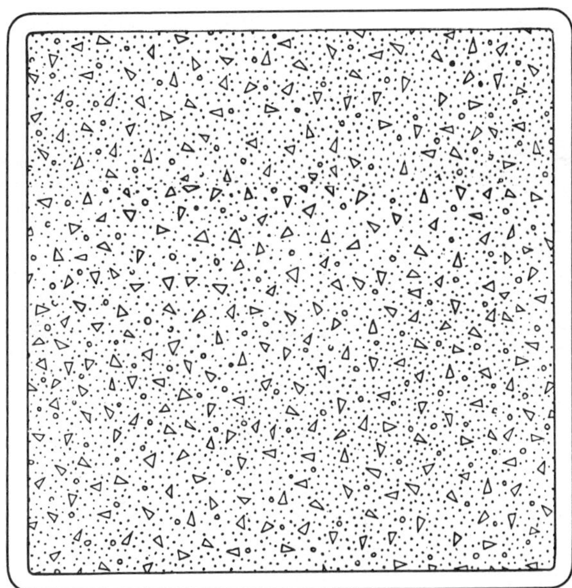

FIGURE 10.11 Concrete-filled structural tube.

A 10-in. tube, then, must have a minimum thickness of 0.2 in., so our $\frac{3}{8}$-inch tube is fine.

Plot two points only, P_0 and the balance point. Assume that all the steel is at yield.

Ultimate Axial Capacity

$$P_0 = A_sF_y + 0.85f'_cA_c$$
$$= [13.8(36) + (0.85)(3.0)(85.5)]$$
$$P_0 = 714.8 \text{ kips}$$

The Balance Point

$$\frac{x}{d} = \frac{0.003}{F_y/E_s + 0.003}$$

Use $d = 10 - 0.375 - 0.188 = 9.44$ in.

$$x = \frac{0.003(9.44)}{F_y/E_s + 0.003} = 6.68 \text{ in.}$$

$$a = 0.85x = 5.68 \text{ in.}$$

Internal Forces

$$C_c = 0.85f'_cab = 0.85(3.0)(5.68)(9.25) = 134 \text{ kips}$$

$$C_{s3} = 2\left(\frac{3}{8}\right)\left(\frac{9.25}{2}\right)(36) = 124.9 \text{ kips}$$

$$C_{s4} = (10)\left(\frac{3}{8}\right)(36) = 126 \text{ kips}$$

$$T_1 = (10)\left(\frac{3}{8}\right)(36) = 126 \text{ kips}$$

$$T_2 = 2\left(\frac{3}{8}\right)\left(\frac{9.25}{2}\right)(36) \times 124.9 \text{ kips}$$

$$P_b = C_c + C_{s3} + C_{s4} - T_1 - T_2$$
$$= 134 + 124.9 + 126 - 126 - 124.9$$
$$P_b = 134 \text{ kips}$$

Taking moments about the centerline,

$$Pe = M = 134(2.16) + 124.9(2.31) + 126(4.81) + 126(4.81) + 124.9(2.31)$$
$$= 289.4 + 288.5 + 606.1 + 606.1 + 288.5$$
$$M = 2078.6 \text{ kip-in.}$$
$$M = 173.2 \text{ kip-ft}$$

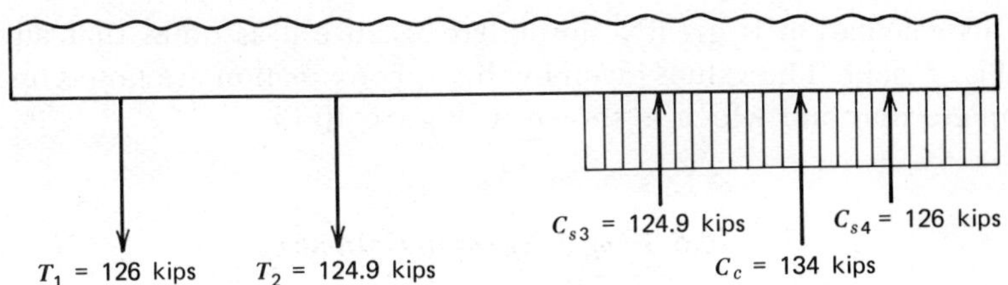

FIGURE 10.12

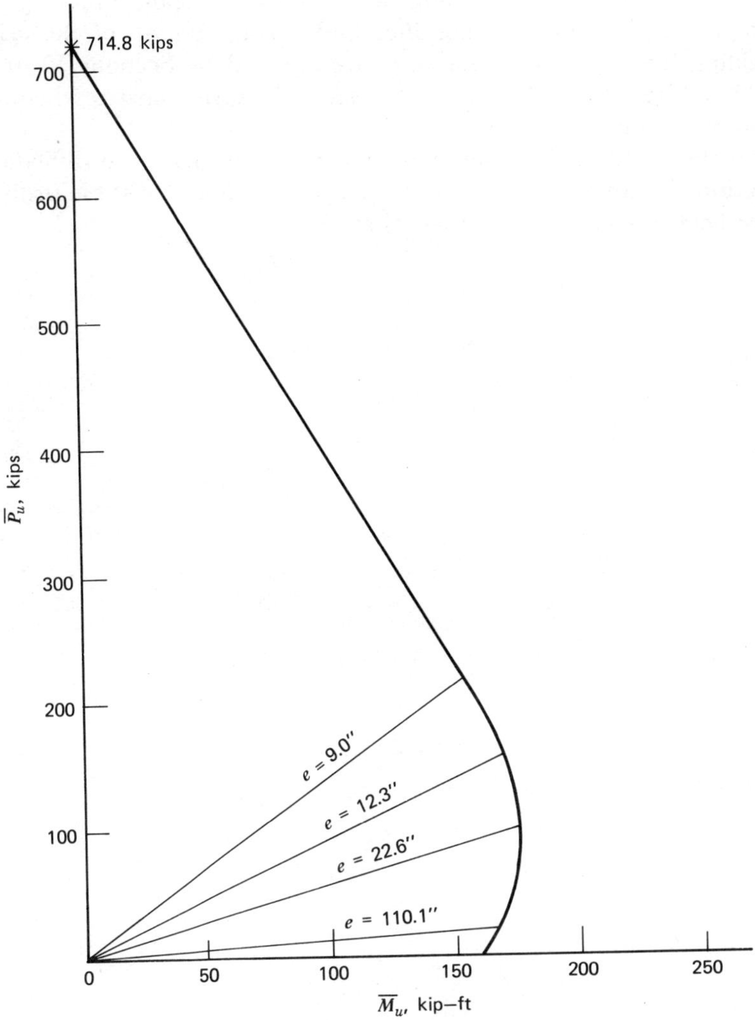

FIGURE 10.13 Interaction diagram.

This calculation is greatly simplified because it assumes that all the steel is at yield. The values given by this approximation are noted on the complete diagram which is shown in Figure 10.13.

10.6 Length Effect in Columns

Length effect in columns means different things to the designer and the builder. To the designer, a long slender column means a more precise design using a moment magnifier and a computation of the critical buckling load. Slenderness effects are covered by Sections 10.10 and 10.11 of the ACI code and are illustrated by standard textbooks on reinforced concrete design.

To the builder, the long slender column means more difficulty in erection, forming, and casting of the columns. Some of these problems have been discussed in a previous section.

11

Wood-Plywood Systems

Wood and plywood have long been used together in residential and other light construction. Plywood laid over wood floor joists produces a smooth solid base for any type of floor finishing. The plywood goes down fast and saves money. One good carpenter can subfloor a medium-sized house in one day.

The plywood and wood joist form the T-shaped section which is typical of composite beams. However, the nail is a poor shear connector. The nail holds the plywood down, but not securely enough to provide composite action. The new APA* glued floor system is based on elastomeric adhesives that bond the structural plywood firmly and permanently to the floor joists.

Wood and plywood are also combined to form a lightweight joist. This joist nicely fills a gap in the market. It covers spans and loads too great for light wood framing and yet not large enough for the usual bar joist to take over. In this joist, the flanges are adhesively bonded to the web.

11.1 Adhesive Bonding

There is a natural attraction to the concept of a continuous shear connection between the components of a composite member, rather than having discrete shear connectors which are accompanied by stress concentrations.

One of the characteristics of the synthetic resins generally is their high shear strength, which suggests their use as a connecting device. Elas-

*American Plywood Association.

tomeric adhesives formulated from the synthetic resins are generally used in the wood systems. It is important to note that the adhesive, while contributing stiffness and strength to the floor system, must have enough resilience to relieve impact, shrinking and other secondary stresses. The adhesive should also be of a mastic type, rather than a liquid type, so that it has some gap-filling capability. Even finely sanded woods have some irregularity across the bond line due to the rings of spring wood and summer wood. The adhesive should also form a good bond under relatively low pressures. The adhesive should have a reasonable "open time" for assembly, but it should set at a rapid rate soon after so that if necessary, the assembly can be moved without damage to the bond. There are several excellent elastomeric construction adhesives which are available in standard caulking cartridges for field gluing. APA states that field gluing virtually eliminates the squeaky floors which result from shrinking lumber. Also, field glued floor systems have better creep resistance than the nailed systems.

11.2 The Wood-Plywood Composite Floor

Just as in any other composite T beam, the flooring and joist act in unison to carry the load. The neutral axis shifts upward from the midheight of the joist, close to the intersection of the plywood and the joist.

The shear strength of the adhesive is the critical factor in the design of an adhesive-bonded composite floor system. It must equal or exceed the strengths of the materials being joined. When the shear strength of the adhesive exceeds the allowable rolling shear stress for plywood, it is considered to have met this requirement.

11.2.1 Effective Portions of the Plywood Panel

In tension, compression, and bending, only the plies that have their grain parallel to the direction of stress are effective. The plies at right angles are subject to stress across the grain, and therefore, they make little contribution to the strength of the panel. If the plywood sheets are oriented at 45° to the joists, the entire strength of the panel can be considered. However, this is so impractical in terms of construction time that it is seldom seen.

The floor system is designed as a series of T-beam assemblies or a continuous, one-sided, stressed skin panel.

The effective width of the T section depends on the joist spacing, the total thickness, t, of the plywood, and the thickness, t_1, of the plies whose grain is parallel to the joist. First compute the basic spacing, B, from

$$B = 36t\sqrt{\frac{t}{t_1}}$$

If the basic spacing is greater than the clear distance between joists, the full center-to-center joist spacing can be used as the effective width.

The basic spacing also determines the stress coefficient which is used to find the allowable stress from Figure 11.1.

When dealing with the wood-plywood system, one new term arises, "rolling shear," which is related to the usual horizontal shear. The rolling shear property of the section simply considers only those plies of the plywood which are parallel to the joist. Figure 11.2 is a flow chart for the proportioning of a floor system.

Example 11.1 Proportion a composite floor system for a residence with a span of 12 ft. The live load is 40 psf.

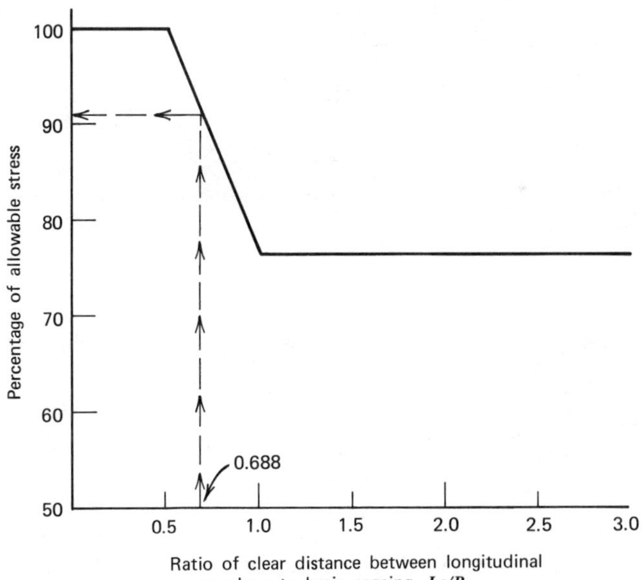

FIGURE 11.1 Variation of allowable stress with spacing of framing members.

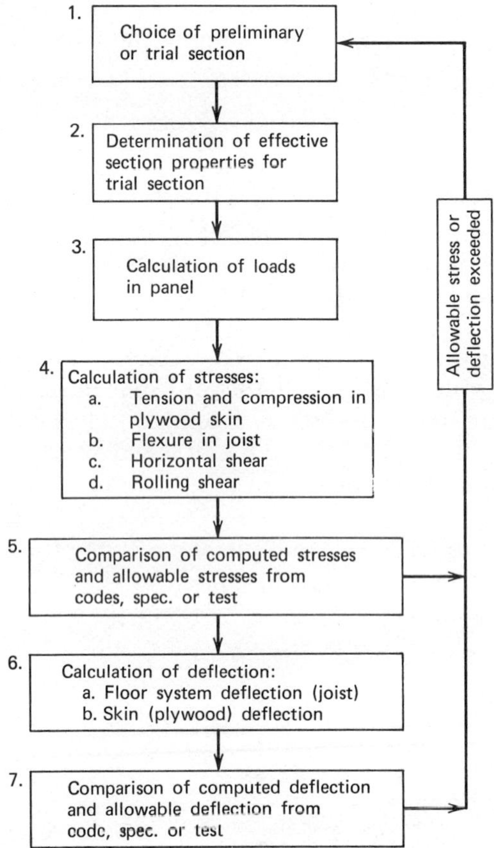

FIGURE 11.2 Flowchart for glued wood floor system.

Preliminary Trial Section

Based on experience, try 2×8 joists and $\frac{1}{2}$-in. plywood subflooring. Joist spacing is 16 in. c/c.

DL $\frac{1}{2}$-in. plywood	1.5 psf	
CEILING AND FIXTURES	1.5 psf	
LL	40 psf	
	43.0 psf	
LOAD ON T SECTION	$43 \times 1.33 = 57.2$ ppf	
JOIST WEIGHT	$= 2.5$ ppf	
	$= 59.7$ ppf	say 60 ppf

END SHEAR (V)

$$V = \frac{60 \times 12}{2} = 360 \text{ lb}$$

$$M = \frac{wL^2}{8} = \frac{60 \times (12)^2}{8} = 1080 \text{ ft-lb}$$

Section Properties

Figure 11.3 shows the preliminary panel section. Actual size of the nominal 2×8 is $1\frac{1}{2} \times 7\frac{1}{4}$ in.

Figure 11.4 shows the effective panel section. Placing the plywood with the outer plies parallel to the joist gives an effective thickness of the plywood sheet of 0.300 in.

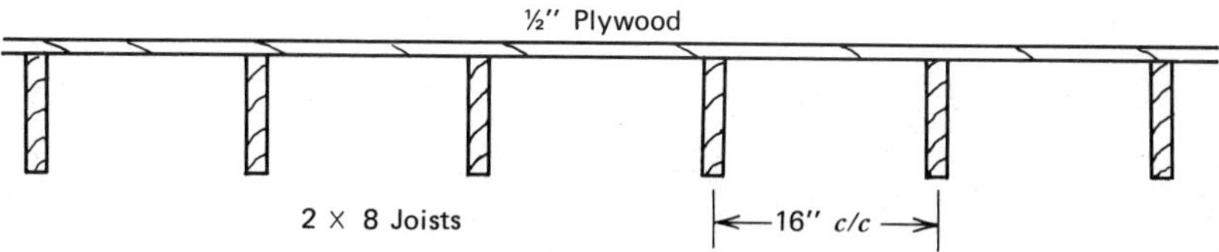

FIGURE 11.3 Wood joist floor system.

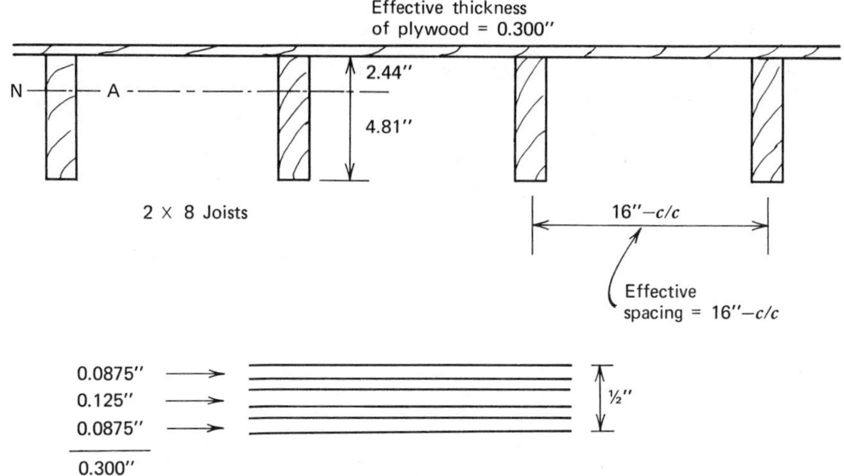

FIGURE 11.4 Effective section.

Note that this is not in strict accordance with the recommendations for the APA glued floor system. The APA recommends placing the 4 × 8-ft plywood sheet across the joists, rather than having the outside plies parallel to the joist. Their recommendation makes good sense. The rolling shear along the joists is more than adequate for five-ply subflooring, and placing three of the five plies across the joists gives greater transverse stiffness and smaller deflections in the plywood between joists.

BASIC SPACING $$B = 36t\sqrt{\frac{t}{t_{\text{parallel}}}}$$

$$= 36(0.5)\sqrt{\frac{0.5}{0.3}}$$

$$= 23.24 \text{ in.}$$

CLEAR SPACING $16 - 1.5 = 14.5$ in.

CLEAR SPACING $< B$

∴ c/c spacing of joists is the effective width (effective width = 16 in.)

STRESS COEFFICIENTS $= \dfrac{\text{Effective width } L_s}{B} = 0.688$

From Figure 11.1, the percentage of allowable stress is 91%. This percentage is concerned with the buckling of the plywood and is applied to the allowable compressive stress in the plywood.

Section	A	y	Ay
Plywood	(0.3)(16) = 4.8	7.5	36.0
Joist	10.875	3.63	39.42
	$\Sigma A = 15.68$		$\Sigma AY = 75.42$

$$y_b = \frac{75.42}{15.68} = 4.81 \text{ in.}$$

Moment of Inertia: I gross $= I_o + AY^2$

Section	I_o	A	y	y^2	Ay^2	$I_o + AY^2$
Plywood	0.04	4.8	2.94	8.64	41.5	41.5
Joist	47.64	10.88	1.19	1.40	15.3	62.9
					I gross =	104.4 in.^{4}

Because the elastomeric adhesive does not provide complete composite interaction, the moment of inertia is modified by a construction factor, C $(= 0.96)$.

$$\text{Effective } I = 0.96(104.4) = 100.2 \text{ in.}^4$$

Rolling Shear Property

This property is similar to the factor Q in the statics formula VQ/I. However, the critical rolling shear plane lies within the plywood between the inner parallel ply and the adjacent perpendicular ply when the plywood has its face grain parallel with the framing members (see Fig. 11.5). When the face grain of the plywood is perpendicular to the framing members, as in the APA glued floor system, the critical rolling shear plane lies between the inner perpendicular plane and the framing member.

$$Q_R = A_1 y_1$$

$$y_1 = 2.44 + (0.5 - 0.087 - 0.05) = 2.80 \text{ in.}$$

$$Q_R = (16 \times 0.3125)(2.80) = 14 \text{ in.}^3$$

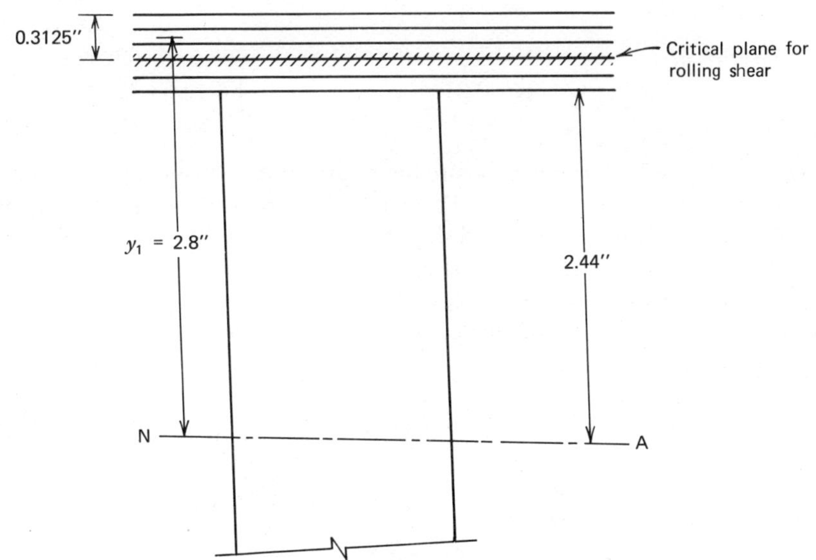

FIGURE 11.5 Rolling shear properties.

Stresses

Compression in plywood:

$$f = \frac{Mc}{I} = \frac{(1080)(12)(2.94)}{100.2} = 380 \text{ psi}$$

Tension in bottom of joist:

$$f = \frac{1080(12)(4.81)}{100.2} = 622 \text{ psi}$$

Rolling shear stress:

$$f_r = \frac{VQ_R}{I_b} = \frac{(360)(14)}{(100.2)(1.5)} = 33.5 \text{ psi}$$

Allowable stresses:

FLEXURE IN JOIST 1800 psi
COMPRESSION IN PLYWOOD 1375 psi
ROLLING SHEAR 53 psi

Modifications:

DRY SERVICE CONDITIONS, FACTOR = 1
PLYWOOD BUCKLING, FACTOR = 0.91
CONTINUOUS DURATION OF LOAD, FACTOR = 0.90

Modified allowables:

FLEXURE IN JOIST	$0.9 \times 1800 = 1620 \text{ psi} > 622 \text{ psi}$	OK
COMPRESSION IN PLYWOOD	$0.91 \times 0.90 \times 1375 = 1126 \text{ psi} > 380 \text{ psi}$	OK
ROLLING SHEAR	$0.9 \times 53 = 47.7 \text{ psi} > 33.5 \text{ psi}$	OK

Deflection of both the composite beam and the plywood skin between the joists should be checked. The deflection of the composite beam simply uses the familiar deflection formula for a uniformly distributed load (see Chapter 8). However, the plywood skin uses a modification of that formula, using a different constant and a moment of inertia based on only those plies running across the joists. See Table 4.14 of the AITC *Timber Construction Manual.*

ALLOWABLE Δ $\quad \dfrac{\text{clear span}}{360} = \dfrac{16\text{–}1.5}{360} = 0.04 \text{ in.}$

ACTUAL Δ $\quad \dfrac{5}{768}\dfrac{wL^4}{EI_{11}}$ (use load in lb/in.)

$$= \frac{5(4.4)(14.5)^4}{768(1.6 \times 10^6)(0.016)} = 0.04 \text{ in.}$$

This floor system is quite obviously overdesigned. The span could be lengthened or a shallower joist could be used. The spacing of the joists should not be widened because it would cause too much deflection of the plywood subfloor between joists.

11.3 The APA Glued Floor System

The APA glued floor system uses elastomeric adhesives to bond the plywood to the wood joists. Nails are used to keep the glue in intimate contact. The recommended nailing is 6*d* deformed shank nails at 12 in., center to center. If the deformed shank nails are unavailable, 8*d* commons at the same spacing can be used. Tongue-and-groove plywood is recommended. Thicknesses specified are $\frac{1}{2}$, $\frac{5}{8}$, or $\frac{3}{4}$ in., depending on joist spacing. If tongue-and-groove plywood is unavailable, square-edge plywood can be used if 2×4 blocking is used under the panel edge joints between joists. The 4×8 plywood sheets are placed with the face grain perpendicular to the joists. End joints of the panels should be staggered, and the blocking, when used, should also be glued. The adhesive can be laid by a bead from a caulking gun. Tests have shown that the intermittent nailing forces the members together so that the result is a good bond line. The edge grooves of the tongue-and-groove plywood should not be set tightly. A $\frac{1}{16}$-in. space should be left at the joints, and the tongue joints should also be lightly glued.

Complete information on the glued floor system, including load and span tables, are available from the American Plywood Association.

11.3.1 Stressed Skin Panels

The glued floor system is essentially a one-sided, stressed skin panel. The two-sided, stressed skin panels may be used for floor and/or ceiling panels or wall panels. The panels are available with a decorative finish on one side.

The panels are usually made in 4 ft widths because this is the normal width of the plywood panels. Panels 8 ft long are common, although longer panels can be used. If longer panels are used, scarf joints are the preferred method of joining plywood sheets. Butt joints can be used if prespliced plywood skin is used. If butt joints are used, and not prespliced, a row of blocking should run across the panel at the joint. Both butting ends should be glued to the blocking.

A cross section of the panel shows that the longitudinal framing members and the two skins form a series of built-up composite I beams with lumber webs and plywood flanges. As with other I shapes, most of the stress is carried by the flanges, in this case the plywood. The longitudinal lumber stringers are usually on 16-in. centers. If the plywood skins are continuous and no concentrated loads are being applied to the panel, internal blocking between the longitudinal members is recommended. In most of the better panels, one row of cross blocking is included at the midlength of a 4×8-ft panel.

Headers should be placed across the ends of the panels to provide stability. The headers should be provided with vent holes to prevent condensation within the panel. Insulation may be provided in the panels if desired.

Design of the panels is somewhat similar to the design of the composite floor system. The complete design procedure is outlined in Section 4 of the AITC *Timber Construction Manual.*

11.4 The Wood-Plywood Joist

The wood-plywood joist is actually a small wood plate girder. This joist uses structural lumber for the flanges and plywood for the web. In this sense, it is the converse of the APA glued floor system which uses the plywood as the flange.

The wood-plywood joist is a viable concept because it provides spans in a range beyond the normal use of wood floor, joist construction. However, these very lengths cause a problem. Straight, true, knot-free pieces in 30 to 40 ft lengths are scarce. One solution is to splice shorter pieces of flange together. Structurally, however, flange splices in the moment-carrying area are poor practice, so a logical step might be to fabricate the flanges as a series of thin plates with the splices staggered. One company has carried out this laminating process, using very thin layers of veneer so that the splices and natural defects in the wood are scattered. The result is the Micro-Lam* flange which bonds thin layers of veneer under heat and pressure into a structural unit. Although this flange consists of several layers, it is considered a structural lumber, rather than plywood, because all the plies are parallel to the longitudinal axis of the flange. Flanges for individual joists are ripped from wider billets to a standard size of $1\frac{1}{2} \times 2\frac{5}{16}$ in. In the commercially made joists, the flanges are one standard size and web depths vary.

Webs are fitted into a groove in the flange and adhesively bonded. The webs are $\frac{3}{8}$-in. structural plywood with the face grains vertical. The

*Trademark, Trus-Joist Corporation.

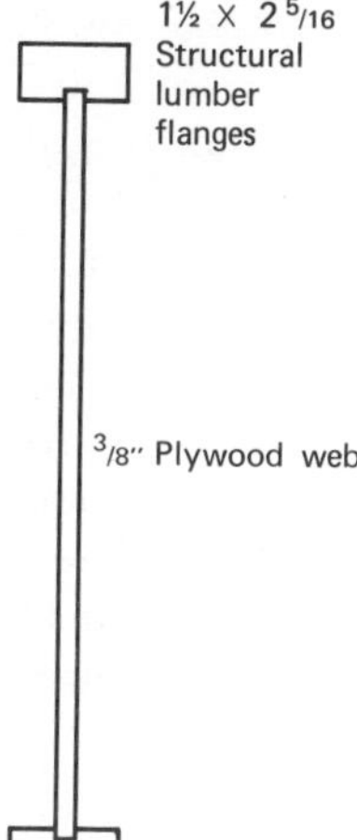

FIGURE 11.6 Wood-plywood joist.

FIGURE 11.7 Wood-plywood joists in place. (Courtesy of Trus-Joist Corporation).

plywood sections are butt jointed and glued to form a continuous web (Fig. 11.7).

These are solid web joists. However, holes can be cut in the web to accommodate ducts, piping, and conduit, provided structural common sense is used. Holes should not be cut through the web close to a support where the shear is high. As with any joist, flanges should not be cut.

The joists are generally designed for uniformly distributed loads, and bridging is used to tie the joists together. If there are concentrated loads within the span, stiffeners should be used at the points of concentrated load. Stiffeners should be used in pairs and nailing should go through the web and into the opposite stiffener.

As with other joist systems, these joists should be doubled under heavy concentrated loads.

The wood-plywood joists are commercially available in depths of 10 to 20 in. for use in spans up to 40 ft, depending on load span and load. Tables are available from the manufacturers.

References

11.1 *Timber Construction Manual,* 2nd ed., American Institute of Timber Construction, Wiley, New York, 1974.

11.2 *APA Glued Floor System,* American Plywood Association, Tacoma, Washington, 1973.

11.3 Igwebuike, G. C., "An Investigation of an Elastoneric Adhesive for Composite Construction," M.S. Thesis, University of Cincinnati, Ohio, 1973.

11.4 *Wood Handbook* (No. 72), Forest Products Laboratory, Madison, Wisconsin, U.S. Department of Agriculture.

11.5 Bergin, E. G., *Glued Bond Failure,* Forest Products Research Bulletin No. P102, Ottawa, Canada.

11.6 Cook, J. P., *Construction Sealants and Adhesives,* Wiley, New York, 1970.

11.7 *Evaluating Adhesives for Building Construction,* USDA Forest Services, FPL No. 172, 1972.

11.8 *Timber Design and Construction Handbook,* Timber Engineering Company, McGraw-Hill, New York, 1956.

11.9 Northcott et al., *Can the Service Life of Bond be Predicted?* ASTM, STP No. 401, 1966.

11.10 Bickerman, J. J., "Strength of Adhesive Joints," Journal of American Chemical Society, September 1972.

12

Steel Deck and Joist Construction

The light-gage, metal deck composite system is probably the fastest growing type of construction on the market today. The metal deck composites offer savings of up to 15 to 20% in the cost of steel framing. They also provide a minimum floor thickness which means more savings. The metal decks also offer a reduction in field labor costs. There is no wooden formwork to be built and usually no shoring is required. However, one of the biggest attractions of this type of composite construction is to the owner. The decks make possible a wide range of services throughout the entire usable floor area. The channels or cells in the metal deck can be used for electrical and telephone wiring, air conditioning, and in some configurations, recessed lighting for the floor below. Some of the available configurations of the metal deck are shown in Figure 12.1. These are only a few of the wide range of shapes available.

12.1 Composite Deck

The composite floor uses the metal deck in conjunction with a concrete slab to form the composite floor. Special embossments on the decking act as shear connectors. The connectors not only resist the horizontal shear, they also provide vertical shear resistance against uplift of the slab (see Fig. 12.2). In the composite deck, the composite action takes place parallel to the direction of the cells in the steel decking and no stud shear connectors are used. Usually a minimum of $2\frac{1}{2}$ in. of concrete is specified, and the concrete is provided with wire mesh or other reinforcement to account for the shrinkage of the concrete.

Economical span lengths vary for the different types of steel sections.

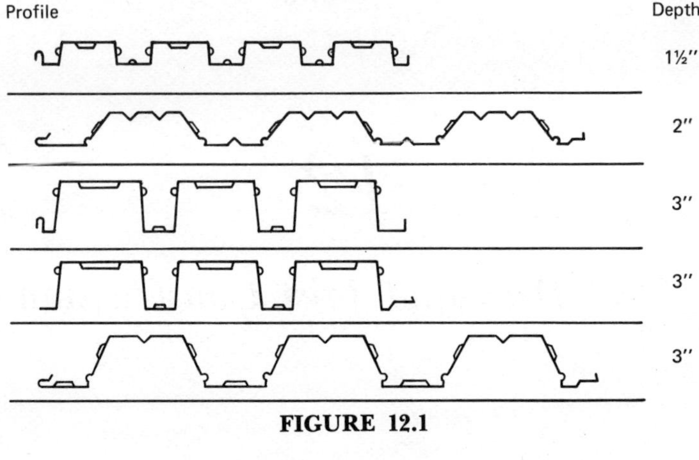

FIGURE 12.1

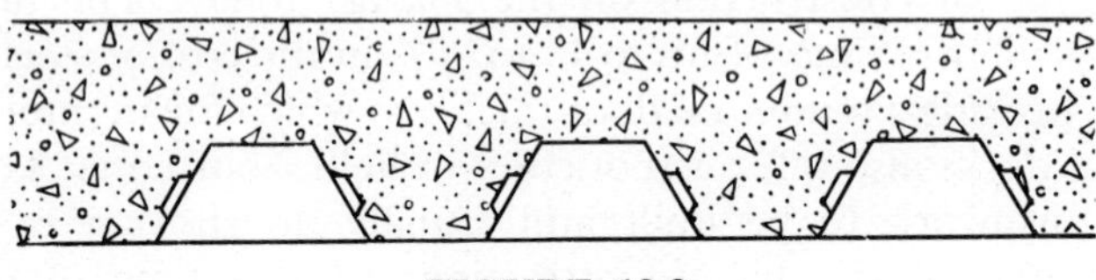

FIGURE 12.2

Spans in the 10 to 15-ft range are common with the 3-in. deep metal sections, giving a total floor thickness of $5\frac{1}{2}$ to 6 in. However, with certain configurations, spans of up to 30 ft can be achieved with a metal deck section $7\frac{1}{2}$ in. deep and a total floor thickness of 10 to 12 in.

12.2 Composite Beams

In the metal deck composite beam system, the corrugations or cells can run either parallel or perpendicular to the supporting steel beams. Stud shear connectors are used to transmit the horizontal shear force (Fig. 12.3). No other type of shear connector is suitable for use with the metal deck.

With modern stud guns, the headed studs are welded directly through the deck to the beam flange. In some cases, additional plug welds are used to tack the decking to the steel beams. These are primarily for construction purposes, but they do provide some slight assistance with the composite action.

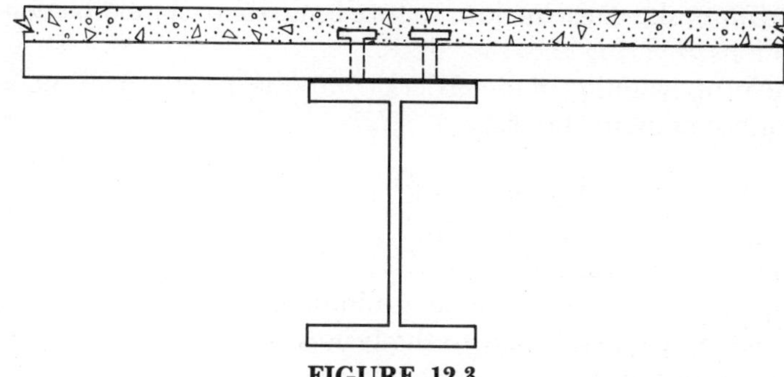

FIGURE 12.3

In the composite beam system, the configuration of the metal decking must permit enough concrete cover around the stud so that the stud can function effectively. Tests have shown that there is a shear cone formed around the stud when specimens are tested to destruction. Overlapping of the shear cones causes a reduction in the individual capacity of each connector when too many studs are placed in one rib trough.

In order to develop the full capacity of the shear connectors, the geometry of the decking should provide a ratio of w/h of at least 1.75 (Fig. 12.4).

The studs used with metal decking should be long enough to extend well into the concrete slab. The stud should be $1\frac{1}{2}$ in. longer than the height of the deck rib.

Research has confirmed the effectiveness of the metal deck composite beam using rib heights up to 2.2 in. Current research is concentrating on the 3-in. deck because this is a practical size which lends itself well to

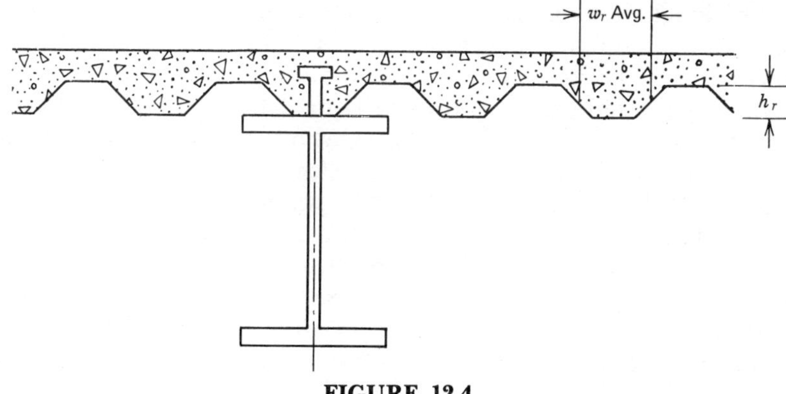

FIGURE 12.4

two-way composite construction. Before long, there should be definitive specifications covering the 3-in. deck.

For the proportioning of metal deck composite beams, AISC makes the following recommendations:

1. Deck rib height (h_r) limited to 2.2 in.
2. Stud diameter should be $\leq \frac{3}{4}$ in.
3. Height of stud connector $= h_r + 1\frac{1}{2}$ in.
4. Use total slab thickness for determining effective width.
5. For decking perpendicular to the beams, use only the concrete area above the decking.
6. Stud capacities are reduced:

$$\text{REDUCTION FACTOR} = 0.6\frac{W_r}{h_r}\left(\frac{H}{h_r} - 1.0\right)$$

This is used for perpendicular decking and for parallel decking when $Wr/h_r < 1.5$.

12.3 Construction

Metal deck composite systems not only offer flexibility in construction and the utilization of floor space, they also cover a wide range of carrying capacities from the light $5\frac{1}{2}$-in. deck reinforced with mesh to the very heavily reinforced sections as shown in Figure 12.5.

The decking is galvanized and can be shop cut, bundled, and marked for a specific bay in a building. If the flooring is shop bundled, all the decking for a particular bay can be hoisted into place at once. Individual decking pieces vary in size, but a typical one might be 3 in. deep by 3 ft wide and can easily be handled and positioned by two men. The edges of the individual decking pieces are matched so that they fit together somewhat like tongue-and-groove pieces. After the decking for a particular bay is hoisted into place, the individual pieces are spread out and the edges fitted together. If there is any field cutting to be done, it should be done before fastening the decking into position. The decking is welded to the steel frame with fusion welds.

Locations for electrical outlets should be planned in advance so that the metal decking can be prepunched to accommodate the outlet boxes. However, the outlet boxes would probably not be installed immediately. Construction sequence would probably have the decking following a story or two behind the structural steel workers. This gives an added feature because the decking, once installed, acts as a safety platform for the steel workers.

FIGURE 12.5

Two stories or so below the decking crew, the electricians and plumbers would be running electrical wiring, and so on through the cells and setting the outlet boxes. These are set so that the service outlets will be flush with the finished floor.

When the outlet boxes are positioned, they should be set at the proper screed level. After the boxes are in place, the reinforcing mesh can be positioned and the slab cast. Most of the metal deck systems are designed for use with 3000-psi concrete. When the slab is cast, an area of about 2 ft in diameter around the boxes should be hand finished, so that when the slab hardens, the entire floor will be flush.

Before the slab is cast, the deck acts as a working platform for the service trades. These trades need materials to work with, and the materials are naturally stacked on the steel deck. Virtually all of the metal deck sections used for composite floors are designed for these construction loads, but the engineer should check the section to be sure.

If the metal decking must be stored at the construction site, it should preferably be stored under cover. If the material is stacked, it should be clear of the ground and provided with enough intermediate wood blocking so that little or no dead load deflection occurs in the stack. If the material is stored outdoors, it should be tilted slightly so that water does not pond in the units.

If the ends of the deck units must be butted together in the building, the joint should be made over a supporting beam and both pieces of the decking should be welded to the beam. Ends of the cells at the exterior of the building should be closed off with coverplates.

12.4 Design of the Composite Deck

The composite deck can be designed using the steel section in tension and the concrete to carry the compression. The geometry of the steel section affects the allowable stress in the steel. The maximum allowable stress is the yield point divided by the factor of safety. This maximum value may have to be reduced if the section contains wide plate elements. Thickness of the plate elements is generally from 12 gage down to 22 gage, depending on the configuration of the section. Table 12.1 relates the standard metal gages to actual thickness for galvanized sheet. These thicknesses are for hot-dipped galvanized sheet. Some of the decking used has only a single layer of "wipe-coat," so that thickness for these units would be slightly less.

Design of these composite decks should conform to the latest edition of the *Specifications for the Design of Light Gage Cold Formed Steel Structural Members* of the American Iron and Steel Institute (AISI).

A design can be performed using the transformed section method of analysis shown in previous chapters. It is generally good practice to limit the span of the composite floors to 32 times the total depth of the floor section.

Generally, a detailed design is not performed by the structural engineer. Section properties and span-load tables are available from the manufacturers of the floor sections. These span-load tables are calculated according to the latest AISI specifications and are verified by certified test

Table 12.1

Gage	Thickness, in.
12	0.1084
14	0.0785
16	0.0635
18	0.0516
20	0.0396
22	0.0336

results on full-scale sections performed by an independent testing agency or university.

If computations are to be made, the usual transformed section method must include some extra considerations. When designing with wide, thin compressive elements, either the allowable stress is reduced or only part of the plate area may be considered as effective. In the composite deck, the compression plates are in contact with the concrete and are close to the neutral axis, so that buckling of these elements is usually no problem.

However, before the slab has hardened, the steel deck section must carry the dead loads and construction loads. Therefore, the steel deck must be checked and the effective width of the compressive plates may be smaller than the actual width (see Reference 12.7).

The composite deck is a shallow section and consequently deflection may be critical under dead and construction loads. The span-load tables from the manufacturer usually include recommendations for the shoring of certain sections and spans to prevent excessive deflection.

12.5 Steel Joist Construction

The open web steel joist is well established as an efficient load-carrying member. The first open web joists were manufactured in the early 1920s. Use of the joists has grown steadily so that these members are now available in six standardized series, in depths to 72 in. and for spans up to 120 ft.

Joist dimensions are standardized and span-load tables are available from the Steel Joist Institute or in the AISC *Manual of Steel Construction.*

Up to this time a composite joist has not been standardized by the Steel Joist Institute. However, research is under way which could lead to standardization of this type of construction.

These joists are typically used at relatively close spacings, and because of this close spacing, a thin slab is used. Wood formwork is usually not used. Corrugated steel sheet, ribbed metal lath, or heavy paper with welded wire mesh are generally used to form the slab. Figure 12.6 shows a typical joist with a 2-in. slab.

The joists are very efficient load-carrying members, but they are flexible. An inspection of the span-load tables for the various joist series shows that for most practical span lengths, the allowable load is controlled by deflection, rather than by the strength of the joist. The composite slab and joist give increased stiffness, so that deflections are decreased and the full strength of the joist can be utilized. For example, a 20-H5 joist is tabulated as having a capacity of 238 ppf based on strength. However, the

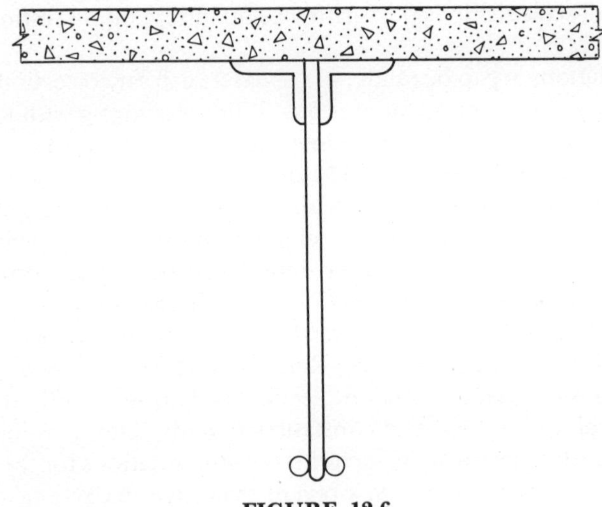

FIGURE 12.6

load, as limited by deflection, is 141 ppf. The same joist, acting compositely with a 2-in. slab, has a capacity of 250 ppf. The stiffer composite joist under the 250-ppf load has a deflection that is less than 1/360 times span. Thus the addition of the composite action increases the effective carrying capacity of the joist from 141 to 250 ppf, an increase of over 75%.

These, of course, are only calculated values. The joist is a very slender member, and further research will undoubtedly set limitations on the composite joists. Some excellent research has been done, but a great deal more is necessary before a composite joist can be standardized.

References

12.1 Fisher, J. W., *Design of Composite Beams with Formed Metal Deck*, American Institute of Steel Construction, *Engineering Journal*, Vol. 7, No. 3, 1970.

12.2 Robinson, H., *Tests on Composite Beams with Cellular Deck*, Proc. Am. Soc. Civil Eng., Str. Div. 93, 1967.

12.3 *Q-Lock/CB: Composite Beam Technical Data Book*, H. H. Robertson Co., Pittsburgh, Pennsylvania, 1971.

12.4 Dallaire, E. E., *Cellular Steel Floors Mature, Civil Engineering*, Vol. 41, No. 7, July 1971.

12.5 *Standard Specifications and Load Tables*, Steel Joist Institute, Arlington, Virginia, 1975.

12.6 Winter, G., *Cold Formed Light Gage Steel Construction*, Proc. Am. Soc. Civil Eng., Vol. 85 (ST-9), 1959.

12.7 Ostapenko, A., "Light Gage Cold Formed Members," in *Structural Steel Design*, L. S. Beedle, Ed., Ronald Press, New York, 1964.

12.8 Tide, R. H., and Galambos, T. V., "Composite Open Web Steel Joists," *AISC Engineering Journal,* Vol. 7, No. 1, 1970.

12.9 Wang, P. C. and Kaley, D. J., "Composite Action of Concrete Slab and Open Web Joists," *AISC Engineering Journal,* Vol. 4, No. 1, 1967.

12.10 Goble, G. G., "Shear Strength of Thin Flange Composite Sections," *AISC Engineering Journal,* Vol. 5, No. 2, 1968.

13

Innovative Composite Designs

The well-known composite beam which combines a beam and slab with a shear connector produces significant economy in construction. Both labor and material costs have risen sharply over the past. Labor costs, especially for field labor, have gone up even more rapidly than the cost of materials. Architects, engineers, and contractors are being constantly challenged to provide designs and construction methods that trim the cost of both materials and labor.

Factory-fabricated components usually result in a net saving, even after adding in the shipping cost. Standardization of sizes is especially helpful. Also, the precast or prefabricated unit can be made to better tolerances, which means fewer field misfits and a savings of labor cost.

Prefabricated units are not necessarily the finished building components. Reusable form panels designed for easy erection and disassembly can be built at the contractor's yard and trucked to the site. With a little extra thought, forms can be built to utilize standard components, such as full size 4×8 plywood sheets, to minimize cutting operations.

Connections cost money. Connections made in the shop are almost always cheaper than field connections. Once at the site, any connections that can be made on the ground are cheaper than those that must be made several stories up on the building frame. With precast concrete, field connections may require special attention. A method of composite construction that eliminates or simplifies these connections is a big money saver.

Up until the late 1930s, many tall buildings were constructed using a structural slab and then placing utilities, such as wiring and duct work, and topping this with a fill slab. This was a relatively thick, heavy floor system, with several separate field operations. Since that time, designers have more and more realized that the construction industry can no longer afford the luxury of single-use building components. There has been a shift to the more widespread use of the cellular steel deck composite floors and composite beams with steel deck.

In more recent years, hemmed in by rising costs, designers have begun reaching beyond the traditional boundaries to combine materials and methods into innovative composite designs.

13.1 Faulkner Hospital Garage, Jamaica Plain, Massachusetts

Information and illustrations for this composite structure were furnished by the courtesy of Volume Indoor Parking, New York, N.Y., and Seelye, Stevenson, Value & Knecht, New York, N.Y.

This 590 car parking garage is a unique two-way composite design. The building uses a combination of precast, prestressed concrete joists and steel girders. This new construction method, known as the CJ/FP system, used a $3\frac{1}{2}$-in.-thick cast-in-place slab to achieve composite action, both with the concrete joists and the steel girders. The building uses 50-ksi steel for the girders and columns and 12-in.-deep precast, prestressed joints. Figure 13.1 shows the joists in place on the girders. The joists have a 4-ft clear spacing, so that full-size plywood sheets can be used for the slab forms.

FIGURE 13.1

Although the girder spans are fairly long, no shoring is required because the specially designed joists provide lateral support for the compression flange of the girder.

The slab forms are supported on aluminum tube "Formstruts," placed 2 ft on centers, spanning between the joists. These tube struts are a special telescoping design that can easily be released, so that the form supports and the plywood slab-form panels can quickly be cycled for reuse.

Usually with precast joists, the field connection of the end of the joists to the main structural frame requires a great deal of attention to detail. In this structure, the joist ends are simply encased in a haunch which forms an integral part of the composite girder. The girders are 30-in.-deep, wide flange sections and are connected to the columns by standard framing connections.

There are several cost-saving features in this design. The building is engineered for a minimum of field work. Slab forms and haunch encasing the joist ends places the slab 9¾ in. above the girder. The combination of the tubular struts and full-size plywood sheets permits fast erection and stripping of the slab.

13.1.1 Construction Sequence

The building consists of three main bays. The girder spans are 58 ft–6 in. in the outer bays and 62 ft–3 in. in the center bay. Girder spacing is 27 ft–0 in.

The precast, prestressed joists have several special features. The joist is dapped at the end to provide a seat as it rests on the girder. The dap serves several purposes, actually. It lowers the center of gravity of the joist in relation to the supporting girder so that the joist is not easily tipped over after being positioned. The remaining height of the joist above the dap cutout determines the depth of the haunch of the composite girder. Figure 13.2 shows the end of one joist resting on the girder. The joists are provided with small holes near the top at 2 ft, center to center, to receive the end pins of the tubular-form supports. These holes can be seen in Figure 13.3. The tops of the joists are "intentionally roughened" for the shear connection to the cast-in-place slab. The support chairs for the slab mesh also act as shear connectors. The configuration of these mesh supports enables them to act as bond-type connectors as well as shear connectors.

After the steel frame is erected, the joists are lifted into place two at a time. Figure 13.4 shows the preliminary positioning of one joist, and Figure 13.5 shows the final positioning using a specially designed spacing bar for precise fit.

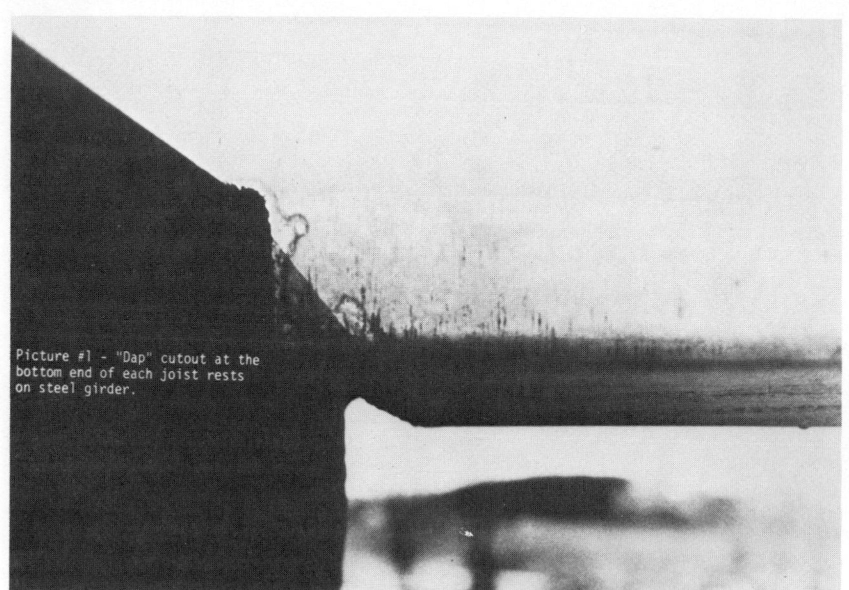

Picture #1 - "Dap" cutout at the bottom end of each joist rests on steel girder.

FIGURE 13.2

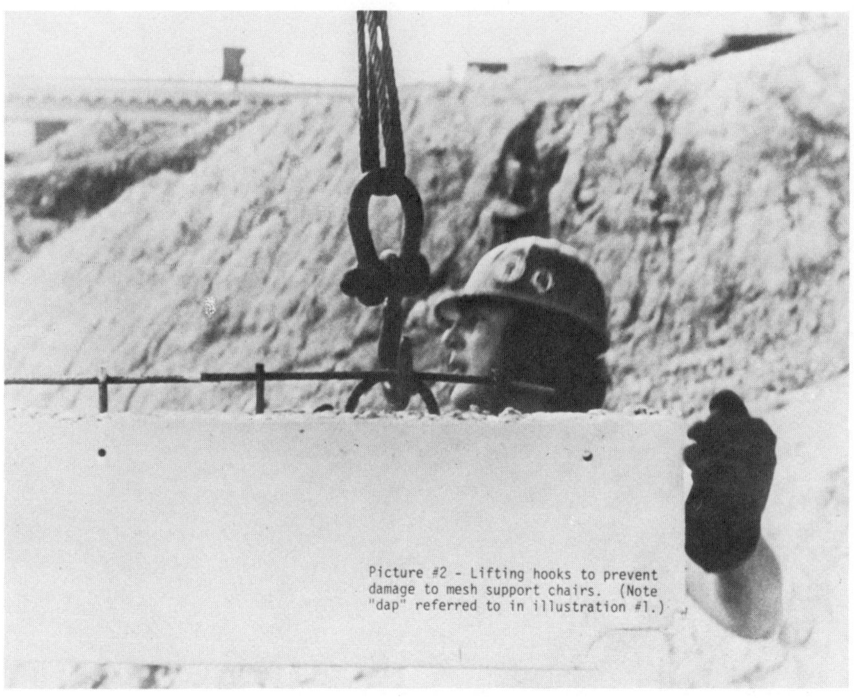

Picture #2 - Lifting hooks to prevent damage to mesh support chairs. (Note "dap" referred to in illustration #1.)

FIGURE 13.3

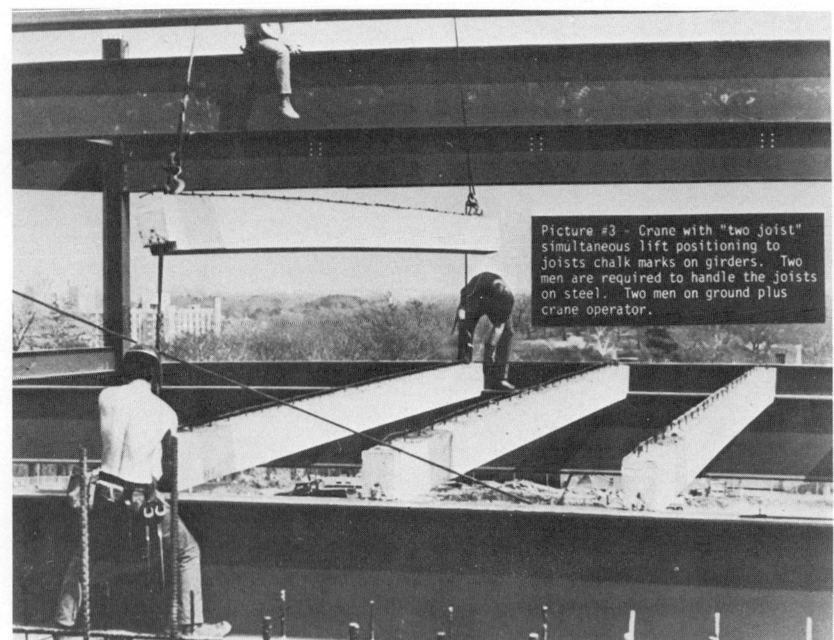

Picture #3 - Crane with "two joist"
simultaneous lift positioning to
joists chalk marks on girders. Two
men are required to handle the joists
on steel. Two men on ground plus
crane operator.

FIGURE 13.4

Picture #4 - Spacing bar facilitates precise positioning.

FIGURE 13.5

This design requires a special 12-in.-long stud to give a positive connection between the girder and the slab. Note in Figure 13.6 that there are, as yet, no studs attached to the girder. In this building, since there are so many joists to be placed, it makes more sense to keep the girder flanges clear so that the joists can be placed quickly. The studs can be installed later.

One of the big problems in unshored, long-span construction is providing lateral support for the compression flange of the steel girder. In this design, the joists provide the lateral support every four ft. As shown in Figure 13.6, every other row of joists is tied together by turnbuckles between the lifting rings on the joist ends. Tightening the turnbuckle pulls the vertical face of the dap in the joist snug against the side of the top flange of the girder. This same picture also shows the haunch form for the composite girder. The tubular slab-form supports and the plywood forms are also shown.

Figure 13.7 shows the neat, uncluttered appearance of the structural system as seen from below.

Picture #5 - Tie alternate rows of joists by placing and tightening turnbuckle units between lifting hooks of adjacent joist ends to brace girders during construction.

FIGURE 13.6

FIGURE 13.7

FIGURE 13.8

This structure was designed to fit the site, which is the side of a hill, so that the higher floors are longer than the lower ones. Figure 13.8 shows the completed structure.

13.2 Fifty-Story Office Building

Information and illustrations for this structure were furnished by the courtesy of Reid & Tarics Associates, San Francisco.

With respect to wind and earthquake loads, this structure is designed as a huge shaft, cantilevered up out of the ground. The central portion of the building is composed of composite columns with a new type of beam-to-column connection. Figure 13.9 shows the portion of the building that uses the composite system. The columns outside the shaded area carry vertical load only. Note that the shaded portion of the structure approximates the shape of the bending moment diagram on this vertical cantilever.

The columns in this building are large-diameter, structural steel pipes filled with concrete. The columns are interconnected with 6-ft-deep plate girders. The girders actually run through the columns. This system helps to control sidesway, greatly simplifies the girder-to-column connections, and helps to provide damping in earthquake areas.

In the normal, steel-framed, tall structure, the columns, especially at the lower story levels, become quite large. Bending induced by sidesway usually makes the columns larger than the axial loads require. The large-scale tubular columns filled with concrete provide the extra stiffness required, so that there is a savings in steel weight. The concrete adds some cost, but the net result is still a slight savings in the cost of the columns.

The biggest savings in this system comes in the number and types of connections. Normally, every interior column requires four beam-to-column connections. The exterior columns require only two or three, depending on location. In this composite frame, the girders go through the columns. The web of the girder is bolted to the column flanges. This is helpful as an erection connection. The concrete inside the column prevents the buckling of the girder web and provides a great deal of the strength of the connection. Not only are the connections simplified, but the number of connections is cut virtually in half.

The fabrication and erection sequence of the beam and column units is shown below. The column sections are fabricated in halves with flanges for assembly and are delivered to the job site. The girders piercing the columns may be enough of a shear connection for the

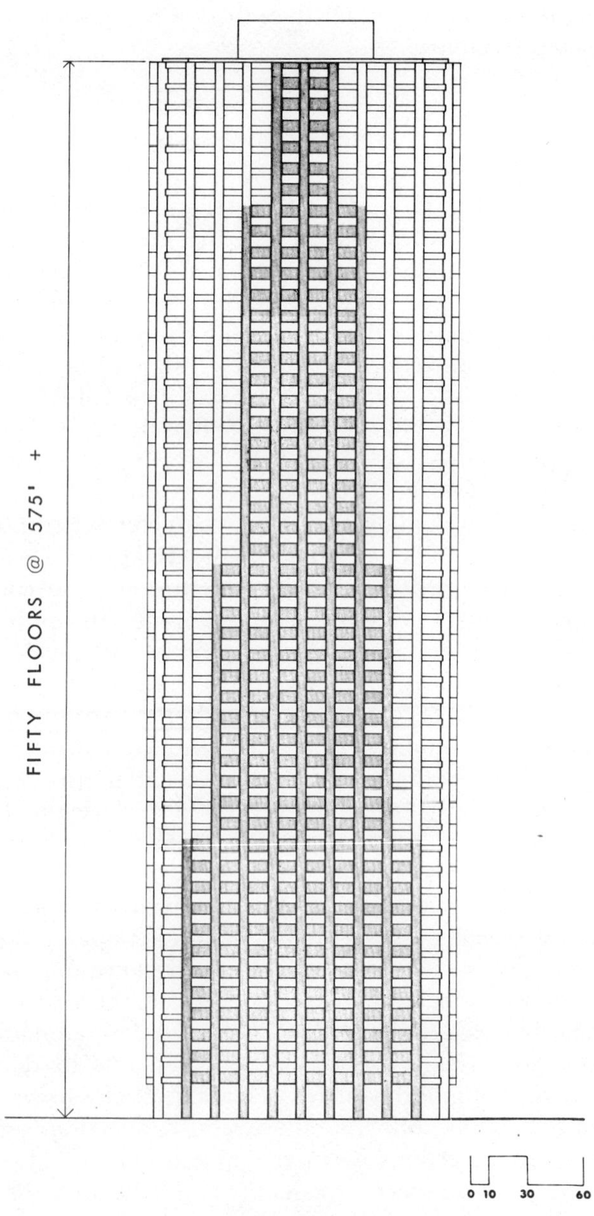

FIGURE 13.9

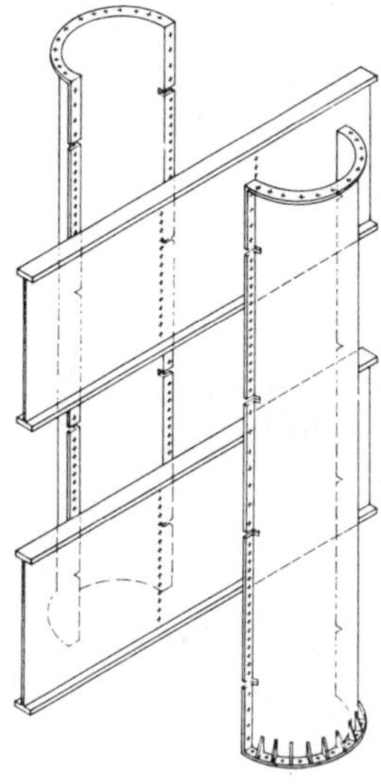

FIGURE 13.10

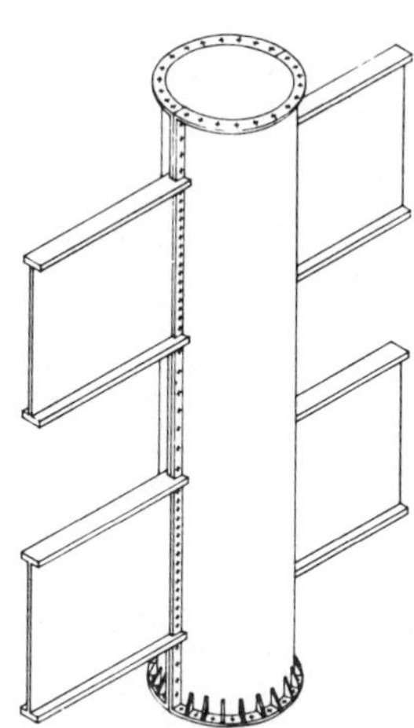

FIGURE 13.11

composite columns. However, since the column sections are in halves, lugs or stud shear connectors can easily be installed, if required by a local code. A unit of one column section with two girder sections is assembled on the ground. All connections are made with high-strength bolts. Note that as many of the bolts as desired can be run through the web of the girder. This unit is then lifted into place and the columns are filled with concrete. If space above the girder permits, it is best to underfill the column by about 1 ft, so that the splices in the steel and the concrete components of the column are not made at the same point. Underfilling will also help to keep the top flange of the column clean, so that the next lift can be erected easily. Dowels should be used to tie the concrete lifts together. Figure 13.12 shows an assembly of two columns and four girders. Girder splices are made at minimum moment points instead of at the column where the stresses and the cost of moment resistant connections are high.

These large-scale tubular columns have not yet been tested for fire resistance. The designers feel that these columns would behave like water-filled columns because of the large amounts of water of crystallization in the concrete mass inside the pipe. If these fire tests are not considered satisfactory, the system can be fire proofed like any other structural steel units.

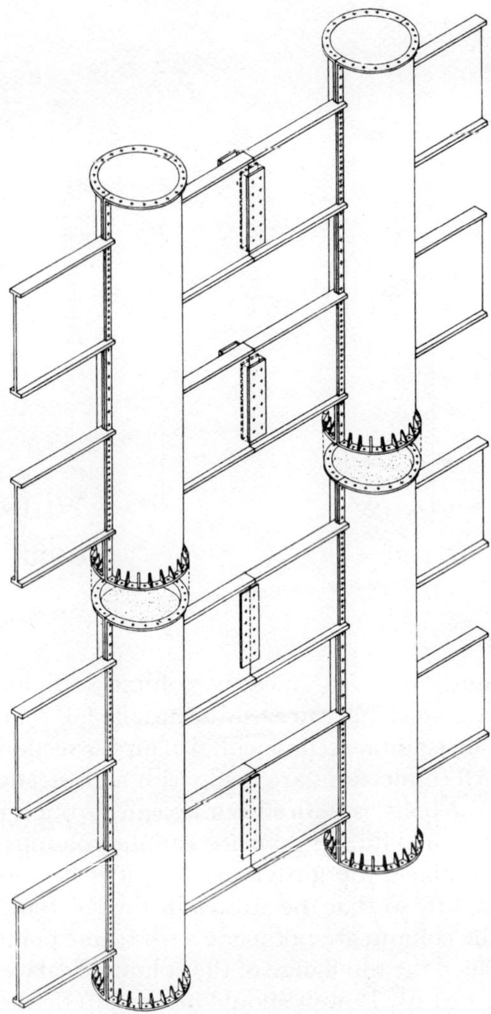

FIGURE 13.12

In sum, this composite beam and column system provides the following:

1. Improved resistance to sidesway.
2. Savings in steel weight.
3. Fewer and simpler highly stressed connections.
4. Fewer field connections.
5. Good clamping and fire proofing characteristics.

13.3 The Duct Beam

This design is a logical extension of the cellular steel deck. For this structure, the main structural elements are box girders which integrate the primary horizontal air duct work into the structural framing.

The box girders can be formed from structural plate so that the top and sides of the girders are in one piece. Bottom plate thickness can be varied to suit the span length. The design uses a steel deck, concrete slab, and stud shear connectors to complete the composite beam.

Another interesting feature of this design is the method of lateral load transfer to the vertical members. Figure 13.14 shows stud shear connectors on both the top and bottom flanges of the beam embedded in the reinforced concrete shear wall.

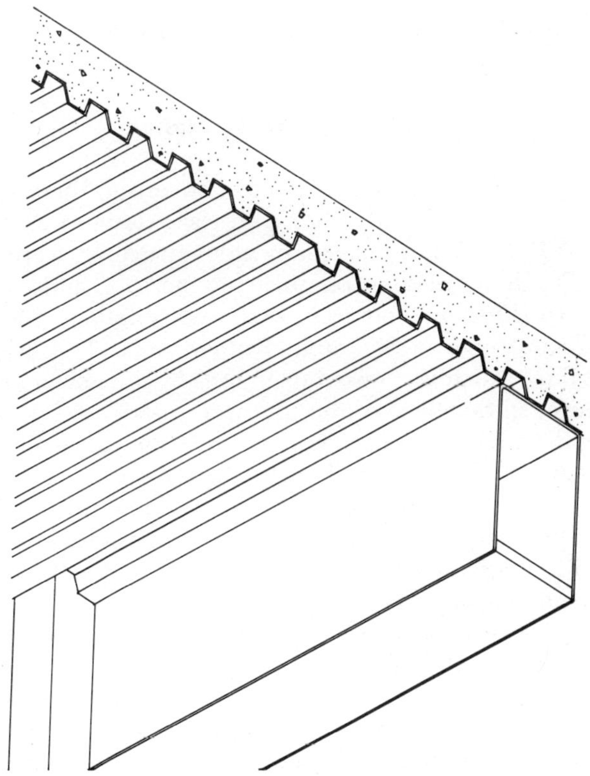

FIGURE 13.13 (Courtesy of Reid and Tarics Associates.)

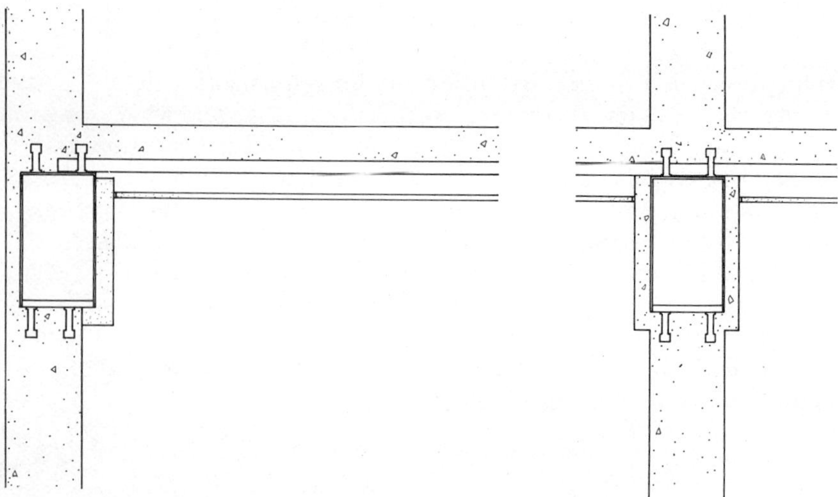

FIGURE 13.14 (Courtesy of Reid and Tarics Associates.)

13.4 Bridge Rehabilitation

The strengthening of older bridges by converting them to composite structures is not new. Shortly after composite design was incorporated into the AASHO specifications, many older structures were stiffened by removing the deck, welding on shear connectors, and casting a new deck. In 1945, this method was used to strengthen the 50-year-old Spruce Street Bridge in Scranton, Pennsylvania. The bridge had an 8-ton load limit which was raised to 15 tons by the new floor system. In this structure, spiral shear connectors were used.

Under today's heavy traffic volumes, it is often not possible to close an entire bridge while a new deck slab cures. Consequently, bridge engineers are devising new methods of attaching precast bridge slab sections to the stringers in order to obtain composite action. The design firm of URS/Madigan Praeger teamed up with the New York State Thruway Authority and the Sika Chemical Company to use this type of repair on a bridge near Amsterdam, New York. Figure 13.15 shows a plan and section of the bridge.

Precast slabs had to be fabricated in sizes which could be quickly hoisted into place. These planks span in the transverse direction. Length of the plank depends on how much of the bridge can be closed. If one-way traffic is being maintained, length of the precast plank is one lane width as shown in Figure 13.16. In the longitudinal direction, a

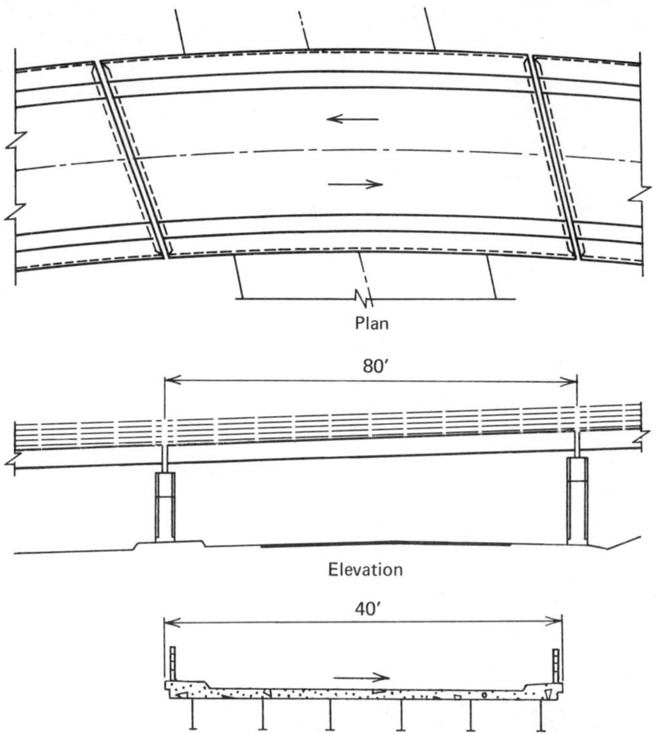

FIGURE 13.15 (Courtesy of Sika Chemical Company.)

FIGURE 13.16 (Courtesy of Sika Chemical Company.)

dimension of about 4 ft provides a convenient size plank for casting and handling. Block-outs were left in the plank to provide the shear connection. The precast planks in this bridge were bedded in epoxy mortar for uniform bearing. The epoxy mortar also provides some extra horizontal shear resistance (Fig. 13.17). In this case, channel shear connectors were welded to the beams inside the block-outs. After the shear connectors were placed, the block-outs were filled with epoxy mortar (Figs. 13.18 and 13.19). Keyways between the adjacent planks were also grouted in with the epoxy mortar (Fig. 13.20). A temporary threaded stud screwed into an insert in the bottom of the plank held the plank in position until the epoxy mortar cured.

This structure was finished off with a bituminous wearing course. This reconstruction method does require careful design and attention to detail, especially at critical points, such as the shear connectors and expansion dams. However, the construction operations are basically

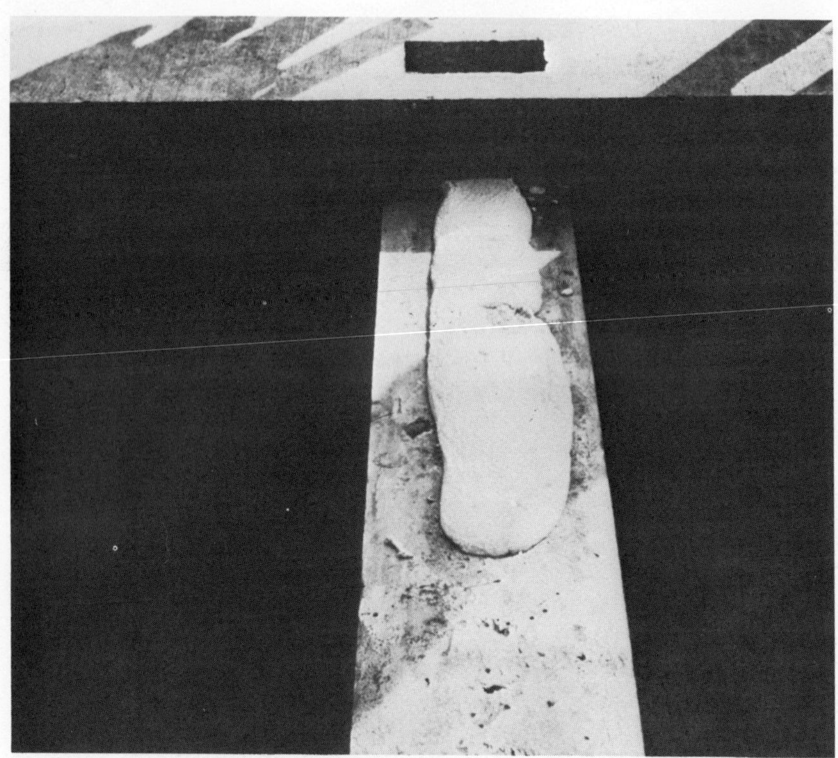

FIGURE 13.17 (Courtesy of Sika Chemical Company.)

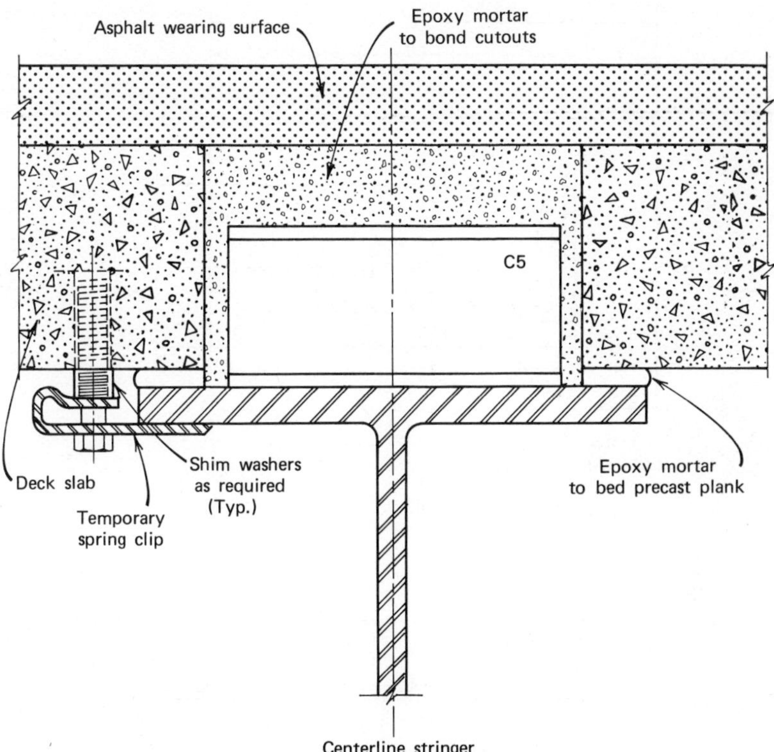

FIGURE 13.18 (Courtesy of Sika Chemical Company.)

simple and straightforward. On the Amsterdam project, all the work, including precasting and installation of the planks, was done by Thruway maintenance personnel.

13.5 Steel-Grid Bridge Floors

The steel-grid floor has a long, successful history of use in bridges. This type of decking has been used on bridges ranging from short-span structures up to such prestigious spans as the Mackinac Straits Bridge and the Verrazano Narrows Bridge.

This type of deck offers many advantages. It has a relatively low dead load. The deck units are factory manufactured into panels which can be speedily erected. The deck when filled with concrete can be designed for

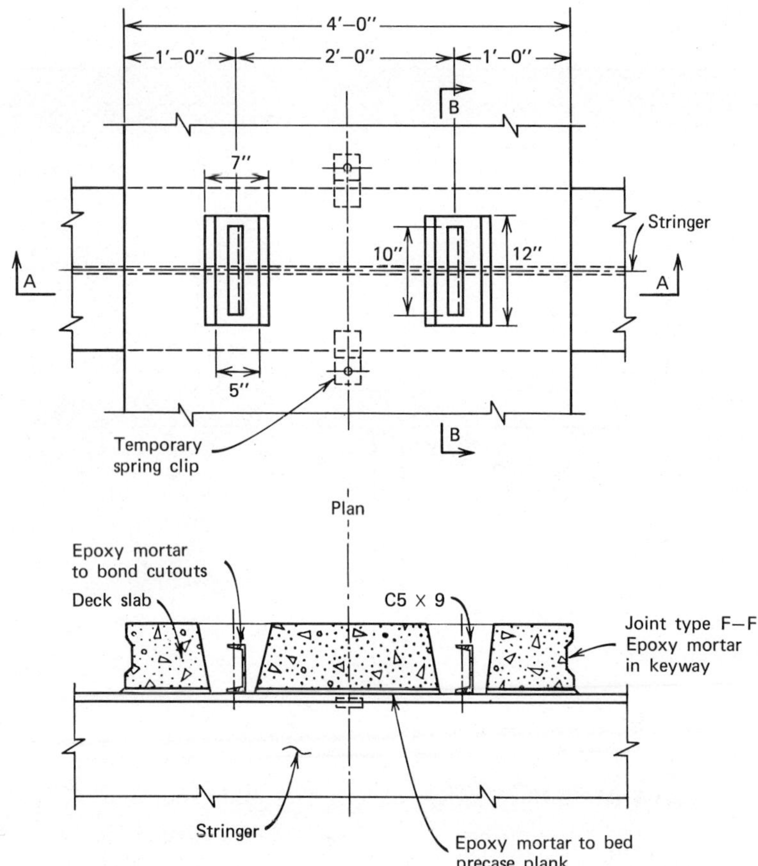

FIGURE 13.19 (Courtesy of Sika Chemical Company.)

composite action with the supporting stringers. The grid units, if desired, can be precast and the filled grid then lifted into position.

The grid floors are available from several manufacturers in a wide variety of configurations to suit the particular structure. The grid floor can be used with the main support members either parallel or transverse to the direction of traffic, depending on the supporting members.

The grid unit consists of main supporting members, which are usually specially rolled I bars or small structural T's, and transverse members to provide stability. The amount of transverse steel is very important. In order to insure that the main support members act together in carrying the load, the amount of transverse steel should be at least 20% of the

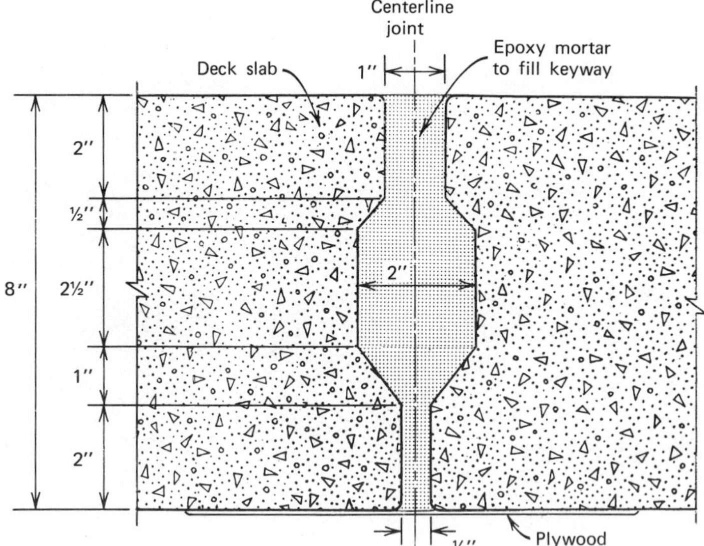

FIGURE 13.20 (Courtesy of Sika Chemical Company.)

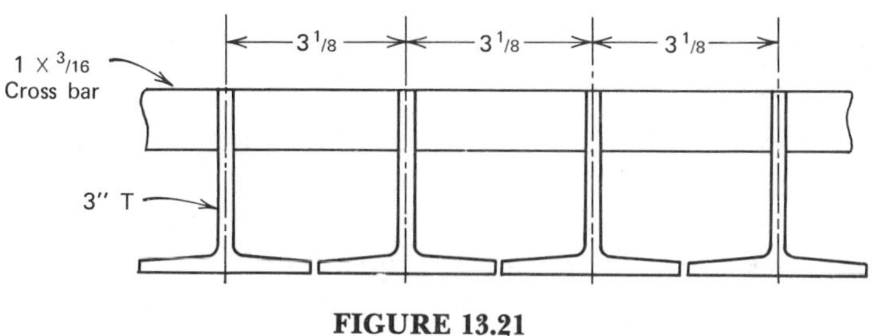

FIGURE 13.21

gross area of the main reinforcement. If the unit is to be concrete filled, a bottom form of 18- or 20-gage sheet may be welded to the unit. Standard formwork can be used under the deck, but it is usually impractical.

The deck units when delivered to the job site must be handled and stored carefully. The units should rest on wood blocking to prevent twisting or warping. If the units have been specified with a shop camber, the blocking should conform to the camber of the units.

When the units are to be placed in the structure, the position of each deck panel should be marked on the supporting stringers. The decking operation can then proceed continuously from one end of the bridge to

FIGURE 13.22 Deck unit with bottom form. (Courtesy of Reliance Steel Products Company.)

the other. Once each unit is swung into place and aligned, it can be tack welded into place. Bridge stringers are seldom truly level and at exactly the same elevation, even in a level bridge. Slight differences in elevation or distortion of the deck panel can be handled in many cases by a large "C" clamp to pull the deck into contact with the stringers. The units are then tack welded to the stringers. As successive panels are placed, a light crane can drive on the previously placed panels. Once the deck units have all been placed, the full amount of design welding can be done.

If the panel is to be concrete filled, the concrete trucks can drive on the grid flooring to place the concrete. Figure 13.23 shows the concrete being placed on a bridge in Morgantown, West Virginia.

If a concrete fill is to be used, the concrete can either be screeded off level with the top of the steel grid or overfilled. If an overfill is to be used, the concrete should be at least $1\frac{3}{4}$ in. thick above the steel. Experience has shown that overfills of 1 in. or less chip off readily. This generally does not impair the structural integrity of the deck, but it has a poor appearance and gives a rough ride.

FIGURE 13.23 Filling the grid deck. (Courtesy of Reliance Steel Products Company.)

In the design of the concrete-filled grid floor for composite action with the supporting stringers, no separate shear connectors are used. The welds attaching the grid floor to the structure furnish the shear connection. Research on the grid decks has suggested that these welds be designed according to channel shear connector fatigue strength, as given in Section 1.7.100 of the AASHTO specifications.

13.6 Epoxy-Bonded Composite Beams

The high shear strength of the epoxy resin adhesives suggests that the epoxies might be suitable shear connectors for composite beams. The big advantage the adhesives have is that they provide a continuous shear connection rather than connection at discrete points. The continuous shear connection provides a continuous stress distribution. On the other hand, the mechanical shear connectors are subject to concentrations of stress.

The disadvantage of adhesive bonding is that the shear strength is provided only at the bond line. The adhesive does not penetrate into the concrete slab, and mortar failures can occur immediately above the bond line. Also, the mechanical shear connector provides a positive clamping action, holding the slab down against the beam.

Nevertheless, an adhesively bonded composite beam in conjunction with a few shear connectors to prevent uplift does show promise.

To be effective as a shear connector, the adhesive needs properties other than strength. The adhesive must be able to set in the presence of water, not be too brittle, should have creep and stress relaxation properties no worse than those of the concrete, and should have the proper viscosity for easy application.

In one research program (13.1) both concrete T beams and steel-concrete composite beams were tested. The control beams were a monolithic T beam and a stud-connected, steel-concrete composite beam. In the case of the concrete-concrete beam, the stem was cast and cured. The top of the stem was then brushed with a stiff brush to remove any laitance. The epoxy bonding agent was then applied by sprayer. Within 4 hr the slab was cast and the beam then cured. The steel beam was prepared in a similar fashion. The epoxy was applied to the clean top flange of the steel beam and the slab cast on to the fresh epoxy.

The results of these tests showed the glued concrete-concrete composite beams to be equal in strength to the monolithic T beam. The steel-concrete composite beam developed $\frac{10}{11}$ of the strength of the stud-connected beam. However, when the steel-concrete beams were severely loaded, then unloaded, and reloaded to failure, they were significantly weaker than the stud-connected beams.

The investigations also noted the tendency for uplift, especially at the ends of the beams, and recommended the use of a few mechanical shear connectors to take the uplift.

13.7 The Bonded-Aggregate Composite Beam

The bonded-aggregate composite beam uses coarse aggregate embedded in a suitable adhesive as a shear connector. This connector should provide a shear strength similar to the value of "intentional roughening" of concrete-concrete composite beams. If the bonded-aggregate proves feasible, considerable construction economy can be achieved. Steel beams could be shop fabricated with the aggregate coating and shipped to the construction site. This would eliminate much of the cost of field installation of shear connectors. The top flange would be protected with heavy paper in shipment to insure a clean bond line. This wrapping would still leave a flat top flange for the steel workers to walk.

The testing program (13.2) for the bonded-aggregate beams included three different types and sizes of aggregates. The adhesive used was a polyester resin. The steel beam flange was sand blasted clean, the

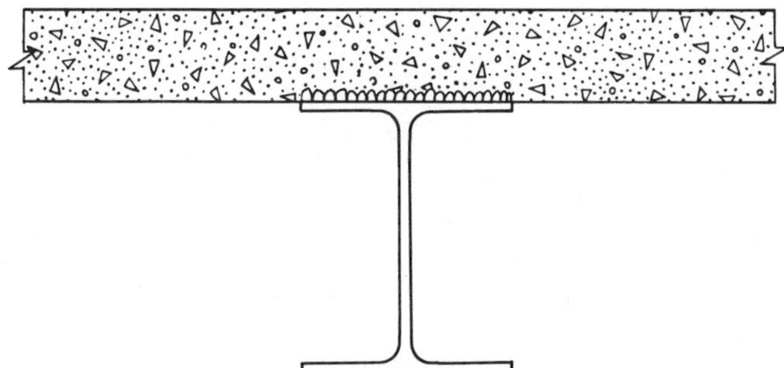

FIGURE 13.24 Bonded-aggregate composite beam.

adhesive applied, and the aggregate sprinkled into the adhesive. After the adhesive was fully cured, the slab was cast and cured.

Three loading tests and one creep test were included in the program. In the loading tests, the strength of the composite beams varied from 1.5 to 3 times as strong as the unbonded control specimens. The creep-loaded specimens showed only minimal increase in deflection after 3 months of loading.

The results of the testing look promising, but they should not be over estimated. In every case, failure of the specimen was preceded by slab uplift at the end of the specimen.

Also in every case, failure occurred in the concrete above the bond line, rather than at the shear connection. In order to take cyclic or impact loads, this type of shear connector would have to be supplemented by a few mechanical shear connectors at the end of the beam.

13.8 Composite Materials

Composite materials are formed by embedding reinforcing fillers, usually fibers, into a matrix material. In a very broad sense, wood and bamboo are composites. They consist of fibers of cellulose embedded in a matrix of lignin which holds the fibers together.

Manufactured composites were first made by the early Assyrians when they embedded straw fibers in a clay to make building bricks.

Today in the construction industry, composites such as hardobard and particle board are finding increasing usage. Hardboards, such as Masonite,* are used for formwork and are veneered for paneling.

*Trademark, Johns-Manville.

Particle boards use a larger wood chip than the hardboards in the matrix of plastic. They are used for shelving, kitchen cabinets, and small furniture. The particle boards are usually veneered. Because of their high wood content, both of these materials can be sawed, drilled, and shaped, much like lumber.

Fiber-reinforced plastics (FRP) and other fiber-reinforced composite materials are now becoming competitive with conventional structural materials. The earlier FRP materials were formed from long strands which were dipped in the matrix. FRP tanks have been formed by a rolling dolly which fed the strand through the matrix plastic and wound it onto the tank form. FRP has a good strength-to-weight ratio, and its use for various structural applications should increase.

Some of the newer composite materials offer a strength-to-weight ratio 4 times as great as high-strength steel. These composites are plastics (usually epoxy) combined with small fibers.

These very small fibers or whiskers have very high strength. Generally, the finer the fibers, the higher is the strength. These fibers have been formed from glass, boron, iron, carbon, and some newer organic fibers. Fiber orientation in the matrix has a great deal to do with the strength. Materials formed with the fibers running in one direction usually have a greater tensile strength than materials with a random fiber orientation. Some of the newer fiber composites have been laminated of layers of unidirectional fibers laid down with the layers crossed, similar to the fabrication of plywood.

These composite materials have been used for specialized purposes, such as trusses for the aerospace industry. The potential for these materials in construction is bright because the materials can be designed for specific uses and the prices are getting more and more competitive.

References

13.1 Miklofsky, H. A., Brown, M. R., and Gonsior, M. J., "Epoxy Bonding Compounds as Shear Connectors for Composite Beams," Physical Research Project No. 13, Bureau of Physical Research, New York State Department of Transportation, 1962.

13.2 Cook, J. P., "The Bonded-Aggregate Composite Beam," Report to Deck-Guard Corporation, Albany, New York, 1967.

13.3 Miklofsky, H. A., Gunsior, M. J., and Santini, J. J., "Further Studies of Epoxy Bonding Compounds," Physical Research Project No. 13, New York State Department of Transportation, 1963.

13.4 Tarics, A. G., "Concrete-Filled Columns for Multistory Construction," *Modern Steel Construction,* Vol. XII, No. 1, 1972.

13.5 *Bridge Floor Systems,* Reliance Steel Products Company, Pittsburgh, Pennsylvania, 1974.

13.6 Eskin, R., "FRP for Structural Applications: The State of the Art and Its Potential for the Future," outline of a thesis, Columbia University, 1972.

13.7 Hooper, I., "A New Structural System for Parking Decks," *Modern Steel Construction,* Vol. XIV, No. 2, AISC, New York, 1974.

Appendices

On the following pages are sections of the appropriate specifications on composite design and construction. These are excerpts, reprinted by permission of the various specifying bodies. The reader is strongly encouraged to obtain, in each case, a copy of the complete specification. Addresses are listed in each instance.

1

Timber-Concrete Composite Construction

SOURCE *Timber Construction Manual*
John Wiley, Publisher

AVAILABLE FROM American Institute of Timber Construction
333 West Hampden Avenue
Englewood, Colorado 80110

or

John Wiley & Sons, Inc.
605 Third Avenue
New York, New York 10016

Composite Beams

9.5 *Composite Beams*

Beams may be designed as composite timber-steel or composite timber-concrete beams.

9.5.1 Composite timber-steel beams shall be so designed that the wood and steel carry the loads in proportion to their relative stiffness.

9.5.2 Composite timber-concrete beams may be either the T-beam type or the slab type. Either type shall be designed in accordance with the methods described in the AITC *Timber Construction Manual.* In the T beam, the effective flange width shall be determined as follows: For composite timber-concrete T beams having the concrete flanges on both sides of the timber beam, the effective flange width of the concrete flange shall not exceed (a) one-fourth of the span length of the beam; (b) 12 times the least thickness of the flange; or (c) the distance center to center of beams. For beams having the flange on one side only, the effective overhanging flange width shall not exceed one-twelfth of the span length of the beam, or six times the least flange thickness, or one-half of the clear distance to the adjacent beam.

9.7.1 RECOMMENDED DEFLECTION LIMITATIONS FOR SAWED AND GLUED LAMINATED BEAMS (SEE TABLE 1)

Table 1

Type of Beam	Use Classification	Live Load Only	Dead + Live Load
Roof beams	Industrial	l/180	l/120
	Commercial and institutional		
	Without plaster ceiling	l/240	l/180
	With plaster ceiling	l/360	l/240
Floor beams	Ordinary usage (construction where walking comfort, minimized plaster cracking, and elimination of objectionable springiness are of prime importance)	l/360	l/240
Bridge beams			
Highway stringers		l/200 to l/300	
Railway stringers		l/300 to l/400	

l is span length in inches for deflection in inches.

9.8 *Camber (see also Appendix 2)*

Provisions shall be made for camber in all glued laminated beams. Practical considerations usually eliminate camber in all glued laminated beams. Practical considerations usually eliminate camber in very short spans and where computed camber is very small. Roof systems require special design considerations to ensure structural adequacy; see paragraph 9.9.

9.8.1 GLUED LAMINATED BEAMS. The amount of camber incorporated at the time of fabrication, and as shown on design drawings, should be at least equal to one and one-half times the calculated dead load deflection. The recommended minimum camber for beams is given in Table 2.

Table 2

Roof beams	$1\frac{1}{2}$ times dead load deflection
Floor beams	$1\frac{1}{2}$ times dead load deflection
Bridge beams	
Long span	2 times dead load deflection
Short span	2 times dead load deflection $+\frac{1}{2}$ live load deflection

Bridge Decks

The selection of decks for timber bridges is determined by density of traffic and economics. Plank decks can be used for light traffic or temporary bridges. Laminated decks can be used for heavier traffic conditions. The design of plank and mechanically laminated decks can be found on pages 4-107 through 4-120. Asphaltic wearing surfaces may be applied on the decking, although this is not usually done for plank decks.

Composite timber-concrete decks are commonly used in timber bridge construction. Composite timber-concrete construction combines timber and concrete in such a manner that the wood is in tension and the concrete is in compression (except at the supports of continuous spans, where negative bending occurs and these stresses are reversed). Composite timber-concrete construction is of two basic types: T beams and slab decks.

T beams consist of timber stringers, which form the stems and concrete slabs which form the flanges of a series of T shapes. Composite beams of this type are usually simple span bridges. Slab decks use as a base for the concrete a mechanically laminated wooden deck made up of planks set on edge, with alternate planks raised 2 in. to form longitudinal grooves. This grooved surface is usually obtained by using planks of two different widths and alternating them in assembly. This composite type is commonly used for continuous span bridges and trestles.

In both types, a means of horizontal shear resistance and of preventing separation are needed at the joint between the two materials. In T beams, resistance to horizontal shear is generally provided by a series of notches, $\frac{1}{2}$ to $\frac{3}{4}$ in. deep, cut into the top of the timber stringer, while nails and spikes partially driven into the top prevent vertical separation of the concrete and timber. Other adequate methods can be used. In slab decks, shear resistance is accomplished either by means of notches, $\frac{1}{2}$ in. deep, cut into the tops of all laminations, by patented, triangular steel plate shear developers driven into precut slots in the channels formed by the raised laminations, or other suitable shear connectors. When the $\frac{1}{2}$-in. notches are used, grooves are milled the full length of both faces of each raised lamination to resist uplift and separation of the wood and concrete. When the steel shear developers are used, nails or spikes are partially driven into the tops of raised laminations to resist separation.

In T-beam design, secondary shearing stresses due to temperature must be considered in designing for horizontal shear resistance. These stresses are induced by the thermal expansion or contraction of the concrete, both of which are resisted by the wood, which is assumed to be unaffected by normal temperature changes. Shear connections for temperature change are neglected in slab deck-type composite construction; however, expansion joints should be provided in the concrete slab.

The concrete slab should be reinforced for temperature stresses. In continuous spans, steel sufficient to develop negative bending stresses is necessary over interior supports.

The dead load of the composite structure is considered to be carried entirely by the timber section. The composite structure carries positive bending moment, and, over interior supports in continuous spans, steel reinforcing and the wood act to resist the negative bending moment.

In designing a composite structure, if it is assumed that the junction between the two materials is without inelastic deformation and has elastic characteristics in keeping with the materials, the structure can be designed by the transformed-area method, that is, by transforming the composite section into an equivalent homogeneous section. This is accomplished by multiplying the concrete width or depth by the ratio of the moduli of elasticity of the materials.

T-Beam Design Procedure

1. Estimate, on the basis of engineering judgment and the use of span tables, the size of timber and thickness of slab to be used.
2. Determine the effective flange width of the concrete slab. The effective width of the slab as the flange of a T beam may not exceed any of the following: (a) one-fourth of the T-beam span; (b) the distance center-to-center of T beams; or (c) 12 times the least thickness of the slab.

   For beams with a flange on one side of the stem only, the effective flange width may not exceed one-twelfth of the beam span, or one-half of the center-to-center distance to the adjacent beam, or six times the least slab thickness.

3. Compute the transformed width of the concrete flange by multiplying the effective width of the flange by the ratio of the moduli of elasticity of the two materials, E_c/E_w.
4. Determine the location of the neutral axis and the moment of inertia, I_t, of the transformed section.
5. Determine applied load and dead load bending moments for the span.
6. Check the extreme fiber stress in bending against the allowable unit stresses for the wood and concrete. The actual extreme fiber stress in bending in the concrete, f_{bc}, is calculated from the formula

$$f_{bc} = \frac{M_A}{S_t}$$

where M_A = applied load bending moment, in.-lb;
S_t = transformed area section modulus, in.^3

The actual extreme fiber stress in bending in the wood, f_b, is determined from the formula

$$f_b = \frac{M_A}{S_t} + \frac{M_D}{S_w}$$

where M_D = dead load bending moment, in.-lb;
S_w = section modulus of the net timber section, in.^3

Impact should be considered in the concrete, but not in the wood.

7. Check the actual horizontal shear stress, f_v, at the timber-concrete junction by the formula

$$f_v = \frac{VxA}{I_t b}$$

where V = total vertical shear, lb;
A = area of the transformed section above or below the plane at which shear is being determined, in.^2;
x = distance of center of gravity of area A from the neutral axis, in.;
I_t = transformed section moment of inertia, in.^4;
b = width of timber section, in.

Since only the net area remaining after the top of the stringer has been notched is effective in shear resistance, the permissible f_v value equals the allowable unit shear stress for the species used multiplied by the ratio of the area notched to the total top surface area of the stringer. *Note*: The top of the stringer need not be notched if properly designed shear connectors are used.

8. The f_v value as determined in step 7 neglects the temperature-change shear connections. The number of connections required, N, may be determined from the formula

$$N = \frac{A_c f_c}{p}$$

where A_c = area of concrete flange considered to be involved by the restraining timber stem, in.^{2} (this may be assumed to be one-third of the total concrete flange area);
f_c = unit stress in concrete induced by temperature change in the range selected, psi (this equals the product of coefficient of expansion of the concrete, the change in temperature from that at the time of construction, and the modulus of elasticity of the concrete, E_c);
p = the value of each shear connection, lb (this may be determined on the basis of allowable shear and bearing values of concrete and wood for notches in the top surface of the stringer; the allowable loads for common fastenings; or tests for special devices).

In order to combine temperature shear requirements and load shear requirements, the formulas in steps 7 and 8 can be rewritten in terms of pounds per inch and combined in the form

$$f_v b = \frac{VxA}{I_t} + \frac{A_c f_c}{12L}$$

where L = span length, ft.

Slab Deck Design Procedure

1. Compute the dead load of the estimated composite section and of the necessary construction loads occurring during placing of concrete and before curing.
2. Compute the bending moments for dead load as for a simple span, and, in the case of continuous-span slab decks, apply the appropriate factor for wood subdeck from Table 4.18 to obtain positive and negative moments.
3. Determine the location of the neutral axis. The distance from the neutral axis to the lower extreme fiber is critical because applied loads add substantially to the lower extreme fiber stress. Compute the moment of inertia of the wood subdeck at midspan and at interior supports of continuous spans. Assume that only two-thirds of the laminations are continuous over supports.

Table 4.18 Maximum Continuous-Span Bending Moments for Slab Decks[a]

	Uniform Dead Load Moments				Applied Load Moments			
	Wood Subdeck		Composite Slab		Concentrated Load		Uniform Load	
	Pos.	Neg.	Pos.	Neg.	Pos.	Neg.	Pos.	Neg.
Span	Percentage of Simple-Span Bending Moments							
Interior	50	50	55	45	75	25	75	55
End	70	60	70	60	85	30	85	65
Two span[b]	65	70	60	75	85	30	80	75

[a] FROM: *Standard Specifications for Highway Bridges*, adopted by The American Association of State Highway Officials, Tenth Edition, 1969.
[b] Continuous beam with two equal spans.

4. From steps 2 and 3, determine the extreme fiber stress in bending in the wood resulting from load at midspan and at interior supports in continuous spans.
5. Compute the bending moments due to uniform dead load after curing and due to concentrated or uniform applied loads as for a simple span. Apply the appropriate constants from Table 4.18 to continuous spans. Concentrated loads are distributed over 5 ft for calculating moments, accounting for equal wheel loads in adjacent lanes.
6. Determine the location of the neutral axis and the distances to the extreme fibers of the wood, concrete, and reinforcing steel. Compute the moment of inertia of the composite section at midspan and over interior supports of continuous spans.
7. Using the M, c, and I values from steps 5 and 6, determine the extreme fiber bending stress in the wood at midspan (tension), in the wood at interior supports of continuous spans (compression), in the concrete at midspan (compression), and in the steel at interior supports of continuous spans (tension). The concrete and steel unit stresses thus determined must be multiplied by the moduli of elasticity ratios, E_c/E_e or E_s/E_w, respectively, to determine actual unit stresses in the transformed concrete and steel areas. These values may not exceed the allowable unit stresses for the materials. Allow-

ance for impact must be added to static moments in determining steel and concrete stresses, but impact is neglected for wood.

The following modulus-of-elasticity ratios should be used:

$E_c/E_w = 1$ for slab decks in which the net concrete thickness above the wood is less than one-half of the overall depth of the composite section;

$E_c/E_w = 2$ for slab decks in which the net concrete thickness above the wood is equal to or greater than one-half of the overall depth of the composite section (the use of a net concrete thickness equal to or greater than one-half of the overall depth of the composite section has little or no advantage for most highway structures);

$E_s/E_w = 18.75$ for Douglas fir and Southern pine lumber.

8. The unit stress in the wood as determined in step 7 plus that determined in step 4 is the total unit stress in the wood at midspan and at the interior supports of continuous spans. This total value may not exceed the allowable unit stress for the lumber grade and species used.
9. Compute the maximum vertical shear adjacent to supports, neglecting the shear due to the dead load of the slab and due to uniform or concentrated loads within a distance of three times the slab depth from the supports. Concentrated moving loads are placed at three times the slab depth or at one-quarter the span from the support, whichever is less, and are distributed over 4 ft for calculating shear.
10. Using the vertical shear from step 9, compute the horizontal shear at middepth of the top channels formed by raised laminations. Use the transformed concrete-wood section for areas of positive moment and the transformed steel-wood section for areas of negative moment. This value may not exceed the allowable horizontal shear stress for the lumber grade used.
11. Compute the spacing of shear developers from step 10, or check the adequacy of notches based on the allowable shear for the concrete or the wood. For a dap length equal to one-half of the dap spacing, the concrete shear will probably control, and thus it may be desirable to make the dap length about one-third of the dap spacing to balance the allowable shear more nearly. If shear developers are employed, the required spacing at the support is generally used as far as the quarter-point of the span and is then increased uniformly to the spacing required (not to exceed 24 in.) by possible vertical shear near midspan.

2

Steel-Concrete Composite Construction

SOURCE — *Manual of Steel Construction*
AVAILABLE FROM — American Institute of Steel Construction
101 Park Avenue
New York, New York 10017

1.11 ***Composite Construction***

1.11.1 *Definition*

Composite construction shall consist of steel beams or girders supporting a reinforced concrete slab, so interconnected that the beam and slab act together to resist bending. When the slab extends on both sides of the beam, the effective width of the concrete flange shall be taken as not more than one-fourth of the span of the beam, and its effective projection beyond the edge of the beam shall not be taken as more than one-half of the clear distance to the adjacent beam, nor more than eight times the slab thickness. When the slab is present on only one side of the beam, the effective width of the concrete flange (projection beyond the beam) shall be taken as not more than one-twelfth of the beam span, nor six times its thickness, nor one-half of the clear distance to the adjacent beam.

Beams totally encased 2 in. or more on their sides and soffit in concrete cast integrally with the slab may be assumed to be interconnected to the concrete by natural bond, without additional anchorage, provided the top of the beam is at least $1\frac{1}{2}$ in. below the top and 2 in. above the bottom of the slab, and provided that the encasement has adequate mesh or other reinforcing steel throughout the whole depth and across the soffit of the beam to prevent spalling of the concrete. When shear connectors are provided in accordance with Section 1.11.4, encasement of the beam to achieve composite action is not required.

1.11.2 *Design Assumptions*

1.11.2.1 Encased beams shall be proportioned to support unassisted all dead loads applied prior to the hardening of the concrete (unless these loads are supported temporarily on shoring) and, acting in conjunction with the slab, to support all dead and live loads applied after hardening of the concrete, without exceeding a computed bending stress of $0.66F_y$, where F_y is the yield stress of the steel beam. The bending stress produced by loads after the concrete has hardened shall be computed on the basis of the section properties of the composite section. Concrete tension stresses shall be neglected. Alternatively, the steel beam alone may be proportioned to resist unassisted the positive moment produced by all loads, live and dead, using a bending stress equal to $0.76F_y$, in which case temporary shoring is not required.

1.11.2.2 When shear connectors are used in accordance with Section 1.11.4, the composite section shall be proportioned to support all of the loads without exceeding the allowable stress prescribed in Section 1.5.1.4, even when the steel section is not shored during construction.

Reinforcement parallel to the beam within the effective width of the slab, when anchored in accordance with the provisions of the applicable code, may be included in computing the properties of composite sections subject to negative bending moment, provided shear connectors are furnished in accordance with the requirements of Section 1.11.4. The section properties of the composite section shall be computed in accordance with the elastic theory. Concrete tension stresses shall be neglected. The compression area of the concrete on the compression side of the neutral axis shall be treated as an equivalent area of steel by dividing it by the modular ratio n.

In cases where it is not feasible or necessary to provide adequate connectors to satisfy the horizontal shear requirements for full composite action, the effective section modulus shall be determined as

$$S_{\text{eff}} = S_s \frac{V'_h}{V_h}(S_{tr} - S_s) \qquad (1.11\text{-}1)$$

where V_h and V'_h are as defined in Section 1.11.4;

S_s = section modulus of the steel beam referred to its bottom flange;

S_{tr} = section modulus of the transformed composite section referred to its bottom flange.

For construction without temporary shoring, the value of the section modulus of the transformed composite section used in stress calculations (referred to the bottom flange of the steel beam) shall not exceed

$$S_{tr} = \left(1.35 + 0.35 \frac{M_L}{M_D}\right) S_s \qquad (1.11\text{-}2)$$

where M_L is the moment caused by loads applied subsequent to the time when the concrete has reached 75% of its required strength; M_D is the moment caused by loads applied prior to this time; and S_s is the section modulus of the steel beam (referred to its bottom flange). The steel beam alone, supporting the loads before the concrete has hardened, shall not be stressed to more than the applicable bending stress given in Section 1.5.1.

The actual section modulus of the transformed composite section shall be used in calculating the concrete flexural compression stress and, for construction without temporary shores, this stress shall be based on loading applied after the concrete has reached 75% of its required strength. The stress in the concrete shall not exceed $0.45f'_c$.

1.11.3 *End Shear*

The web and end connections of the steel beam shall be designed to carry the total dead and live loads.

1.11.4 *Shear Connectors*

Except in the case of encased beams as defined in Section 1.11.1, the entire horizontal shear at the junction of the steel beam and the concrete slab shall be assumed to be transferred by shear connectors welded to the top flange of the beam and embedded in the concrete. For full composite action with concrete subject to flexural compression, the total horizontal shear to be resisted between the point of maximum positive moment and points of zero moment shall be taken as the smaller value using Formulas 1.11-3 and 1.11-4.

$$V_h = \frac{0.85 f'_c A_c}{2} \qquad (1.11\text{-}3)$$

and

$$V_h = \frac{A_s F_y}{2} \qquad (1.11\text{-}4)$$

where f'_c = specified compression strength of concrete;
A_c = actual area of effective concrete flange defined in Section 1.11.1;
A_s = area of steel beam.

In continuous composite beams wherc longitudinal reinforcing steel is considered to act compositely with the steel beam in the negative moment regions, the total horizontal shear to be resisted by shear connectors between an interior support and each adjacent point of contraflexure shall be taken as

$$V_h = \frac{A_{sr}F_{yr}}{2} \tag{1.11-5}$$

where A_{sr} = total area of longitudinal reinforcing steel at the interior support located within the effective flange width specified in Section 1.11.1;
F_{yr} = specified minimum yield stress of the longitudinal reinforcing steel.

For full composite action, the number of connectors resisting the horizontal shear, V_h, each side of the point of maximum moment, shall not be less than that determined by the relationship V_h/q, where q, the allowable shear load for one connector, is given in Table 1.11.4. Working

Table 1.11.4

	Allowable Horizontal Shear Load (q), kips (Applicable only to concrete made with ASTM C33 aggregates)		
	f'_c, kip/in.^{2}		
Connector	3.0	3.5	4.0
$\frac{1}{2}$-in. diam. × 2 in. hooked or headed stud	3.1	5.5	5.9
$\frac{5}{8}$-in. diam. × $2\frac{1}{2}$ in. hooked or headed stud	8.0	8.6	9.2
$\frac{3}{4}$-in. diam. × 3 in. hooked or headed stud	11.5	12.5	13.3
$\frac{7}{8}$-in. diam. × $3\frac{1}{2}$ in. hooked or headed stud	15.6	16.8	18.0
3-in. channel, 4.1 lb.	$4.3w$	$4.7w$	$5.0w$
4-in. channel, 5.4 lb.	$4.6w$	$5.0w$	$5.3w$
5-in. channel, 6.7 lb.	$4.9w$	$5.3w$	$5.6w$

w = length of channel in inches.

values for use with concrete having aggregate not conforming to ASTM C33 and for connector types other than those shown in Table 1.11.4 must be established by a suitable test program.

For partial composite action with concrete subject to flexural compression, the horizontal shear, V'_h, to be used in computing S_{eff} shall be taken as the product of q times the number of connectors furnished between the point of maximum moment and the nearest point of zero moment.

The connectors required on each side of the point of maximum moment in an area of positive bending can be uniformly distributed between that point and adjacent points of zero moment, except that N_2, the number of shear connectors required between any concentrated load in that area and the nearest point of zero moment, shall be not less than that determined by Formula 1.11-6.

$$N_2 = \frac{N_1\left[\dfrac{M\beta}{M_{max}} - 1\right]}{\beta - 1} \qquad (1.11\text{-}6)$$

where M = moment (less than the maximum moment) at a concentrated load point;

N_1 = number of connectors required between point of maximum moment and point of zero moment, determined by the relationship V_h/q or V'_h/q, as applicable;

$\beta = S_{tr}/S_s$ or S_{eff}/S_s, as applicable.

Connectors required in the region of negative bending on a continuous beam may be uniformly distributed between the point of maximum moment and each point of zero moment.

Shear connectors shall have at least 1 in. of concrete cover in all directions. Unless located directly over the web, the diameter of studs shall not be greater than $2\frac{1}{2}$ times the thickness of the flange to which they are welded.

1.11 *Composite Construction—Commentary*

1.11.1 *Definition*

When the dimensions of a concrete slab supported on steel beams are such that the slab can effectively serve as the flange of a composite T beam, and the concrete and steel are adequately tied together so as to act as a unit, the beam can be proportioned on the assumption of composite action.

Two cases are recognized: fully encased steel beams that depend on natural bond for interaction with the concrete, and those with mechanical anchorage to the slab (shear connectors), which do not have to be encased.

1.11.2 *Design Assumptions*

Unless temporary shores are used, beams encased in concrete and interconnected only by means of natural bond must be proportioned to support all of the dead load, unassisted by the concrete, plus the superimposed live load in composite action, without exceeding the allowable bending stress for steel provided in Section 1.5.1.

Because the completely encased steel section is restrained from both local and lateral buckling, an allowable stress of $0.66F_y$, rather than $0.60F_y$, can be applied here. The alternate provision, permitting a stress of $0.76F_y$, to be used in designs where a fully encased beam is proportioned to resist all loads unassisted, reflects a common engineering practice where it is desired to eliminate the calculation of composite section properties.

In keeping with the *Tentative Recommendations for the Design and Construction of Composite Beams and Girders for Buildings,* when shear connectors are used to obtain composite action, this action may be assumed, within certain limits, in proportioning the beam for the moments created by both live and dead loads, even for unshored construction. This liberalization is based on an ultimate strength concept, although the proportioning of the member is based on the clastic section modulus of the transformed cross section.

In order that the maximum bending stress in the steel beam, under service loading, will be well below the level of initial yielding, regardless of the ratio of live load moment to dead load moment, the section modulus of the composite cross section, in tension at the bottom of the beam, for unshored construction, is limited to $(1.35 + 0.35\ M_L/M_D)$ times the section modulus of the bare beam.

On the other hand, the requirement that flexural stress in the concrete slab, due to actual composite action, be computed on the basis of actual transformed section modulus and limited to the generally accepted working stress limit is necessary to avoid excessively conservative slab-to-beam proportions.

Research at Lehigh University has shown that for a given beam and concrete slab, the increase in bending strength intermediate between no composite action and full composite action is directly proportional to the shear resistance developed between the steel and concrete, that is, the number of shear connectors provided between these limits. At times it may not be feasible, nor even necessary, to provide full composite action. Therefore, the specification recognizes two conditions: full and partial composite action.

For the case where the total shear, V'_h, developed between steel and concrete on each side of the point of maximum moment is less than V_h, Formula 1.11-1 can be used to derive an effective section modulus, S_{eff}, having a value less than the section modulus for fully effective composite action, S_{tr}, but more than that of the steel beam alone.

1.11.4 *Shear Connectors*

Based on tests at Lehigh University, and a reexamination of previously published test data reported by a number of investigators, more liberal working values are recommended for various types and sizes of shear connectors than were in use prior to 1961.

Composite beams in which the longitudinal spacing of shear connectors has been varied according to the intensity of statical shear, and duplicate beams where the required number of connectors were uniformly spaced, have exhibited the same ultimate strength, and the same amount of deflection at normal working loads. Only a slight deformation in the concrete and the more heavily stressed shear connectors is needed to redistribute the horizontal shear to other less heavily stressed connectors. The important consideration is that the total number of connectors, on either side of the point of maximum moment, be sufficient to develop the composite action counted on at that point. The provisions of the specification are based on this concept of composite action.

The required shear connectors can generally be spaced uniformly between the points of maximum and zero moments. However, certain loading patterns can produce a condition where closer spacing is required over a part of this distance.

Consider, for example, the case of a uniformly loaded simple beam also required to support two equal concentrated loads, symmetrically disposed about midspan, of such magnitude that the moment at the concentrated loads is only slightly less than the maximum moment at midspan. The number of shear connectors, N_2, required between each end of the beam and the adjacent concentrated load would be only slightly less than the number, N_1, required between each end and midspan.

Formula 1.11-6 is provided as a check to determine whether the number of connectors, N_1, required to develop M_{max} would, if uniformly distributed, provide N_2 connectors between one of the concentrated loads and the nearest point of zero moment. It is based on the requirement that

$$S_{\text{eff}} : S_{tr} = M : M_{\max}$$

where $0 < M < M_{\max}$;

S_{eff} = section modulus corresponding to the minimum amount of partial composite action required at the section subject to the moment M;

$V'_h : V_h = N_2 : N_1$.

In computing the section modulus at points of maximum negative bending, reinforcement parallel to the steel beam and lying within the effective width of slab may be included, provided such reinforcement is properly anchored beyond the region of negative moment. However, enough shear connectors are required to transfer, from the slab to the steel beam, one-half of the ultimate tensile strength of the reinforcement.

The working values for various types of shear connectors are based on a factor of safety of approximately 2.50 against their demonstrated ultimate strength.

The values of q in Table 1.11.4 must not be confused with shear connection values suitable for use when the required number is measured by the parameter VQ/I, where V is the total shear at any given cross section. Such a misuse could result in providing less than half the number required by Formulas 1.11-3, 1.11-4, or 1.11-5.

Stud welds not located directly over the web of a beam tend to tear out of a thin flange before attaining their full shear-resisting capacity. To guard against this contingency, the size of a stud not located over the beam web is limited to $2\frac{1}{2}$ times the flange thickness.

Composite Design for Building Construction

Nomenclature

A_c Actual area of effective concrete flange in composite design

A_s Area of steel beam in composite design

A_{sr} Area of reinforcing steel providing composite action at point of negative moment

E Modulus of elasticity of steel (29,000 kip/in.^{2})

E_c Modulus of elasticity of concrete

F_b Bending stress permitted in the absence of axial force

F_v Allowable shear stress

F_y Specified minimum yield stress of the type of steel being used (kip/in.^{2}). As used in this specification, "yield stress" denotes either the specified minimum yield point (for those steels that have a yield point) or specified minimum yield strength (for those steels that do not have a yield point).

F_{yr} Yield stress of reinforcing steel providing composite action at point of negative moment

I Moment of inertia

I_{tr} Moment of inertia of transformed composite section

K Theoretical coverplate length factor

L Span length (ft)

M Moment (kip/ft)

M_D Moment produced by dead load (loads applied before concrete has hardened)

M_L Moment produced by live load (loads applied after concrete has hardened)

N_1 Number of shear connectors equal to V_h/q or V'_h/q, as applicable

N_2 Number of shear connectors required where closer spacing is needed adjacent to point of zero moment

R Reaction or concentrated transverse load applied to beam or girder (kips)

S_{eff} Effective section modulus for partial composite action

S_j Section modulus of transformed composite cross section, referred to the top of steel beam

S_s Section modulus of steel beam used in composite design, referred to the bottom flange

S_t Section modulus of transformed composite cross section, referred to the top of concrete

S_{tr} Section modulus of steel beam used in composite design, referred to the top flange

V Statical shear on beam (kips)

V_h Total horizontal shear to be resisted by connectors under full composite action (kips)

V'_h Total horizontal shear to be resisted by connectors in providing partial composite action (kips)

b Effective width of concrete slab; actual width of stiffened and unstiffened compression elements

b_f Flange width of rolled beam or plate girder

f_b Computed bending stress

f_c Concrete working stress

f'_c Specified compression strength of concrete

n Modular ratio; equal to E/E_c

q Allowable horizontal shear to be resisted by a shear connector
t_f Flange thickness
β Ratio S_{tr}/S_s or S_{eff}/S_s

General Notes

The AISC specification contains provisions for designing composite steel-concrete beams as follows:

1. For totally encased unshored steel beams not requiring mechanical anchorage (shear connectors), see Sections 1.11.1 and 1.11.2.1.
2. For both shored and unshored beams with mechanically anchored slabs, design of the steel beam is based on the assumption that composite action resists the total design moment (Section 1.11.2.2). In shored construction, flexural stress in the concrete slab due to composite action is determined from the total moment. In unshored construction, flexural stress in the concrete slab due to composite action is determined from moment M_L, produced by loads imposed after the concrete has achieved 75% of its required strength. Shored construction may be used to reduce dead load deflection and must be used if $S_{tr} > \left(1.35 + 0.35\dfrac{M_L}{M_D}\right)S_s$.
3. For partial composite action, see Section 1.11.2.2.
4. For negative moment zones, see Section 1.11.2.2.

General Considerations

1. Composite construction is appropriate for any loading. It is most efficient with heavy loading, relatively long spans, and beams spaced as far apart as permissible.
2. For unshored construction, concrete compressive stress will seldom be critical for the beams listed in the composite beam property tables if a full width slab and $F_y = 36$ ksi steel are used. It is more likely to be critical when a narrow concrete flange or $F_y = 50$-ksi steel is used, and is frequently critical if both $F_y = 50$-ksi steel and a narrow concrete flange are used. Shored construction also increases the concrete stress.
3. Because composite construction usually involves relatively long spans and wide spacing of beams, the specification rule that governs effective slab width is usually the provision limiting the projection

beyond the edge of each beam flange to eight times the slab thickness (see Section 1.11.1).

4. Slab thicknesses of 4 to $5\frac{1}{2}$ in. will be used most often because of fireproofing considerations and because of the wide spacing of beams.
5. Steel and concrete materials of various strengths may be used.

Deflection

A composite beam has much greater stiffness than a noncomposite beam of equal depth, loads, and span length. Deflection of composite beams will usually be about one-third to one-half less than deflection of noncomposite beams. In practice, shallower beams are used and deflections, particularly of the steel section alone under construction loads, should be checked.

Limiting the depth/span ratio may also prevent many deflection problems. The AISC commentary suggests a ratio of $F_y/800$ for fully stressed beams. This yields depth/span ratios as follows:

$$\tfrac{1}{22} \text{ for } F_y = 36 \text{ ksi}$$

$$\tfrac{1}{16} \text{ for } F_y = 50 \text{ ksi}$$

These ratios are offered as simple guidelines; however, the intent of the specification is that a rational calculation of deflections should be made. Such calculations often reveal that smaller depth/span ratios are satisfactory. The depth used in the above ratios is the distance from the top of concrete to the bottom of the steel section.

If it is desired to minimize the transient vibration due to pedestrian traffic when composite beams support large open floor areas free of partitions or other damping sources, the depth/span ratio of the steel beam should not be less than 1/20 for any grade of steel.

Use of Coverplates

Bottom coverplates are an effective means of increasing the strength or reducing the depth of composite beams when deflections are not critical, but they should be used with overall economy in mind. The cost of attaching a $\frac{3}{4}$-in.-thick plate is about the same as for a $\frac{1}{4}$-in. plate.

Two general guidelines for choosing between coverplated and non-coverplated sections of similar capacity are as follows:

1. If the coverplated section would save less than 7 lb/ft, do not use coverplates.
2. If the coverplated section would save more than 12 lb/ft, use coverplates.

Between these limits, minor savings may result from either coverplated or noncoverplated sections. Note that these guidelines may vary from region to region, and should be checked locally.

Other Considerations

The AISC specification provisions for the design of composite beams are based on ultimate load considerations, even though they are presented in terms of working stresses. Because of this, for unshored construction, actual stresses in the tension flange of the steel beam under working load are higher than calculated stresses. Formula 1.11-2 limits the tension flange stress to a value well below yield stress. This same section also provides requirements for limiting the steel beam compression flange stress under construction loading.

Adequate lateral support for the compression flange of the steel section will be provided by the concrete slab after hardening. During construction, however, lateral support must be provided or working stresses must be reduced in accordance with Section 1.5.1.4 of the specification. Steel deck with adequate attachment to the compression flange, or properly constructed concrete forms, will usually provide the necessary lateral support for the type of construction shown in the sketches accompanying the composite beam property tables. For construction using fully encased beams, particular attention should be given to lateral support during construction.

The design of the concrete slab should conform to the ACI Building Code.

Design Aids for Composite Construction

Composite beam tables have been prepared for common conditions encountered in building design and are based on the following:

1. 3.0-ksi concrete ($n = 9$).
2. Two effective flange widths: $16t + b_f$ and $6t + b_f$, where t is the slab thickness and b_f is the compression flange width of the steel section.

3. Concrete slab thicknesses of 4, $4\frac{1}{2}$, 5, and $5\frac{1}{2}$ in.
4. Selected steel beams ranging from 8 to 36 in. in depth. (These will generally be satisfactory for span ranges from 20 to 60 ft for girders, and for longer spans for filler beams.)

For buildings, it will be found that the above conditions will be fully met in most cases. The tabulated effective flange width should be checked against flange width limits based on the given span and beam spacing, in accordance with the provisions of Section 1.11.1.

Explanation of Tables

The tables apply to composite beams having a concrete slab placed directly on a steel beam, where the two elements are connected by stud or channel shear connectors.

Data are included for both coverplated sections and sections without coverplates. These data are applicable to all grades of steel included in Section 1.4 of the AISC specification, except that the tabulated maximum allowable web shear values, V, for coverplated beams apply only to $F_y = 36$-ksi steel.

Selection Tables

Separate composite beam selection tables are given for slab thicknesses of 4, $4\frac{1}{2}$, 5, and $5\frac{1}{2}$ in., for 24 beams with and without coverplates, for full slab width. Beams with partial width slabs are not tabulated.

After the designer has chosen a trial section from the selection tables, the necessary design properties can be obtained from the properties tables to complete the design.

The properties tables also serve as selection tables for 91 beams with no coverplate, since the tabulated beam properties are listed in descending order of transformed section modulus, S_{tr}.

Properties Tables

Two sets of properties tables are provided. The first lists beams with no coverplates and the second lists beams with coverplates.

Values of effective concrete slab width, b, are tabulated for each section for use in the shear connector calculations.

Values of S_{tr}/S_t are tabulated for each section and can be used with footnote 2 of the tables to determine if concrete stress governs.

Values of S_{tr}/S_t at balanced design, below which allowable concrete stress will not be exceeded for shored construction, are listed in footnote 5 of the tables for six strengths of steel and three strengths of concrete.

Concrete stress is much more likely to control in sections with heavy coverplates, high strength steel beams, or partial width slabs, but it should be checked in all cases.

Values of I_{tr} and y_b are given primarily to assist the designer who may wish to check other tabular data or calculate the transformed section modulus at the top of the steel beam, S_j.

Properties Tables for Beams with Coverplates

Tables for beams with coverplates also list average weight per foot, W_3 (maximum weight per foot), K, $12Q/I$, and properties of the steel section alone.

The constant K is a coefficient for determining the theoretical length of the coverplate. It is exact for simply supported beams with uniformly distributed loads. The theoretical cutoff point for any type of loading occurs where the moment is equal to the maximum moment multiplied by the ratio of S_{tr} (noncoverplated) to S_{tr} (coverplated).

The actual required length of coverplate is the theoretical length of the coverplate plus two times the extension length required by the provisions of Section 1.10.4 of the AISC specification. For a simple span with uniformly distributed loading, $L_{cp} = KL$, where L_{cp} is the theoretical length of coverplate in feet and L is span in feet.

The quantity $12Q/I$ is tabulated for use in the formula $F = (12Q/I)M$, where F is total force in kips to be developed by the coverplate end welds, and M is the moment at theoretical cutoff point in kip-ft. For development of the ends of partial length coverplates, the designer's attention is called to the provisions of Section 1.10.4 and Example 3.

The quantity $12Q/I$ may also be multiplied by $V/12$ to determine the horizontal shear in kips per linear inch of beam to be developed by intermediate welds, where V is the vertical shear in kips at the theoretical cutoff point.

Properties of the steel section alone are included for computing construction load stresses and deflections. They are also useful in interpolating for properties of a trial section when the slab width or concrete strength does not conform to the limits of the tables. For convenience, they are repeated for all slab thicknesses.

Properties tables for beams with coverplates are so arranged that interpolation will usually be necessary to find the most economical coverplated section. Interpolation in $\frac{1}{8}$-in. increments of plate thickness will yield practical beam sections. Errors resulting from such interpolation or use of equivalent area coverplates are usually negligible.

Properties Tables for Partial Slab Width

Tables for partial slab widths ($b = 6t + b_f$) are included for the selection of composite beams when the slab is present on only one side (spandrel beams) and to aid in interpolation for narrow flange composite beams (b less than in full slab tables).

Properties from these tables can also be used to estimate long-term creep deflections for full slab composite beams (see discussion on deflection computations).

General Comments

Interpolation between the tabulated slab widths and thicknesses is considered proper. Errors resulting from such interpolation are usually negligible.

By comparing the tabulated I values for flange width $b = 16t + b_f$ and $b = 6t + b_f$, it may be seen that the effect of slight changes of concrete area or b/n ratios is insignificant. Reducing the concrete area 60% results in a reduction of the moment of inertia of only 10 to 15%. For this reason, the tables may be used as a guide for determining trial sections for other concrete strengths, slab thicknesses, and effective widths. More comprehensive tables for a large range of b/n values are available from other sources.

Shear connectors must have a 1-in. minimum cover. Unless located directly over the beam web, stud diameters must not be more than $2\frac{1}{2}$ times the flange thickness (Section 1.11.4).

When shear connectors are used in combination with metal deck, the designer should consult and follow the recommendations of the manufacturer whose deck is being considered. In such cases, research test data should be available to substantiate true composite action and verify design procedures. Here, as with stay-in-place metal forms, a determination of the "efficiency" of the concrete slab must be made, and the connector welding used must provide a proper shear value per connector.

For ready reference, a list of frequently used formulas for composite design follows:

$$n = \frac{E}{E_c} = \text{modular ratio}$$

$$S_{tr} = \frac{12M}{0.66F_y}$$

$$f_b \text{ (steel)} = \frac{12M}{S_{tr}} \text{ (at bottom)}$$

$$= \frac{12M}{S_j} \text{ (at top)}$$

$$f_b \text{ (concrete)} = \frac{12M}{nS_t} \text{ (at top of slab)}$$

$$\text{Maximum } S_{tr} \text{ (for unshored construction)} = \left(1.35 + 0.35\frac{M_L}{M_D}\right)S_s$$

(Formula 1.11-2)

$$V_h = \frac{0.85f'_cA_c}{2} \text{ (for concrete)} \qquad \text{(Formula 1.11-3)}$$

$$= \frac{A_sF_y}{2} \text{ (for steel)} \qquad \text{(Formula 1.11-4)}$$

$$S_{eff} = S_s + \frac{V'_h}{V_h}(S_{tr} - S_s) \qquad \text{(Formula 1.11-1)}$$

$$V'_h = q \times \text{number of connectors furnished}$$

$N_1 = \dfrac{V_h}{q}$; also see stud coefficient tables in the discussion of shear connector computations.

$$N_2 = \frac{N_1 \times \left[\dfrac{M\beta}{M_{max}} - 1\right]}{\beta - 1} \qquad \text{(Formula 1.11-6)}$$

$F = \dfrac{12QM}{I}$ = total horizontal shear force in kips to be developed by welds at the end of coverplate, where M is the moment at cutoff point in kip-ft.

$\dfrac{12Q}{I} \times \dfrac{V}{12}$ = horizontal shear force in kips per linear inch of beam to be developed by intermediate coverplate welds, where V is the vertical shear force at cutoff in kips.

Deflection Computations

Deflections for simple span, uniformly loaded beams at actual loads can be quickly calculated using the formula

$$\Delta = \frac{ML^2}{160Sy}$$

where Δ = deflection, in.;
M = moment, kip-ft;

S = section modulus, in.^{3};
y = distance from bottom of steel section to neutral axis, in.;
L = span length, ft.

For Unshored Beams

DEAD LOAD DEFLECTION $M = M_D$, $S = S_s$, and $y = y_{bs}$

LIVE LOAD DEFLECTION (SHORT TERM) $M = M_L$, $S = S_{tr}$, and $y = y_b$

For Shored Beams

DEAD LOAD DEFLECTION $M = M_D$, $S = S_{tr}$, and $y = y_b$

LIVE LOAD DEFLECTION (SHORT TERM) $M = M_L$, $S = S_{tr}$, and $y = y_b$

If it is desired to consider long-term creep deflection, S_{tr} and y_b should be based on an n value double that shown in the tables. Using S_{tr} and y_b from the partial slab tables for beams with a full slab will give an estimate of this deflection.

End Reactions

If end reactions are not shown on the engineering drawings, they can be calculated using the formula

$$R = \frac{0.33 S_{tr} F_b}{L}$$

This formula is accurate for uniformly loaded, fully stressed, simple span beams and is conservative for most other types of loading.

Shear Connector Computations

The quantity V_h is the total horizontal shear force to be resisted by shear connectors between the points of maximum and zero moment and is computed using Formula 1.11-3 or 1.11-4 in the AISC specification. Using Table 1.11.4 in the specification,* the number of shear connectors required for full composite action is as follows:

$$N_1 = \frac{V_h}{q}$$

*Also see Section 1.11 of Supplement Nos. 1 and 2 to the 1969 AISC Specification for Table 1.11.4A.

The required number of shear connectors thus obtained may be spaced uniformly between the points of maximum and zero moment, except that N_2, the number of shear connectors required between a concentrated load in the area of positive bending and the nearest point of zero moment, must not be less than that determined by Formula 1.11-6. The balance of the shear connectors $(N_1 - N_2)$ may be spaced uniformly between the load point and the point of maximum moment (see Example 4).

Formula 1.11-6 does not apply in negative moment areas of continuous beams where shear connectors are spaced uniformly. In this area, V_h is determined from Formula 1.11-5.

The following table of stud coefficients simplifies the calculation of the value N_1. Coefficients are given for several strengths of concrete and steel combined with q values for stud shear connectors. To utilize the stud coefficient method, find n_1 as the lesser of the values N_c or N_s,

where N_c = number of studs required between point of maximum moment and point of zero moment based on the concrete section
= A_c times the stud coefficient;
N_s = number of studs required between point of maximum moment and point of zero moment based on the steel section
= W_s times the stud coefficient;
A_c = effective concrete flange area in square inches $(b \times t)$;
W_s = weight per foot of the steel section in pounds. (For cover-plated beams, this is the weight per foot at the center of the beam including the plate, not the average weight.)

Stud Coefficients

	For Computing N_s						For Computing N_c		
	$F_y = 36$ ksi			$F_y = 50$ ksi			All Values of F_y		
Stud Size,	f'_c, ksi			f'_c, ksi			f'_c, ksi		
in.	3.0	3.5	4.0	3.0	3.5	4.0	3.0	3.5	4.0
$\frac{1}{2} \times 2$	1.038	0.963	0.897	1.442	1.337	1.246	0.250	0.270	0.288
$\frac{5}{8} \times 2\frac{1}{2}$	0.662	0.616	0.575	0.919	0.855	0.799	0.160	0.173	0.185
$\frac{3}{4} \times 3$	0.461	0.424	0.398	0.639	0.588	0.553	0.111	0.119	0.128
$\frac{7}{8} \times 3\frac{1}{2}$	0.339	0.315	0.294	0.471	0.438	0.408	0.082	0.089	0.094

3

Composite Bridges—AASHTO

SOURCE	Standard Specifications for Highway Bridges
AVAILABLE FROM	American Association of State Highway and Transportation Officials 341 National Press Building Washington, D.C. 20004

The AASHTO specifications cover many types of composite structures, such as steel-concrete I girders and box girders, concrete-concrete T beams and box girders, and hybrid girders. The specifications are quite thorough. Only very brief sections are excerpted here. A full copy of the specifications should be obtained.

Composite Girders

1.7.96 *Composite I Girders—General Specifications*

This section pertains to structures composed of steel girders with concrete slabs connected by shear connectors.

General specifications pertaining to the design of concrete and steel structures shall apply to structures utilizing composite girders, where such specifications are applicable. Composite girders and slabs shall be designed and the stresses computed by the composite moment of inertia method and shall be consistent with the predetermined properties of the various materials used.

The ratio of the moduli of elasticity of steel (29,000,000 psi) to those of concrete of various design strengths shall be as follows:

f'_c = unit ultimate compressive strength of concrete as determined by cylinder tests at the age of 28 days, psi.

n = ratio of modulus of elasticity of steel to that of concrete. The value of n, as a function of the ultimate cylinder strength of concrete, shall be assumed as follows:

$$
\begin{aligned}
f'_c &= 2000\text{–}2400 & n &= 15\\
&= 2500\text{–}2900 & &= 12\\
&= 3000\text{–}3900 & &= 10\\
&= 4000\text{–}4900 & &= 8\\
&= 5000 \text{ or more} & &= 6
\end{aligned}
$$

The effect of creep shall be considered in the design of composite girders that have dead loads acting on the composite section. In such structures, stresses and horizontal shears produced by dead loads acting on the composite section shall be computed for "n" as given above or for this value multiplied by 3, whichever gives the higher stresses and shears.

If concrete with expansive characteristics is used, composite design should be used with caution and provision must be made in the design to accommodate the expansion.

Composite sections should preferably be proportioned so that the neutral axis lies below the top surface of the steel beam. If concrete is on the tension side of the neutral axis, it shall not be considered in computing moments of inertia or resisting moments except for deflection calculations. Mechanical anchorages shall be provided to tie the sections together and to develop stresses on the plane joining the concrete and the steel.

The steel beams, especially if they are not supported by intermediate falsework, shall be investigated for stability during the time the concrete is in place and before it has hardened.

1.7.97 *Shear Connectors*

The mechanical means which are used at the junction of the girder and slab for the purpose of developing the shear resistance necessary to produce composite action shall conform to the specifications of the respective materials as provided in Division II. The shear connectors shall be of types which permit a thorough compaction of the concrete in order to insure that their entire surfaces are in contact with the concrete. They shall be capable of resisting both horizontal and vertical movement between the concrete and the steel.

The capacity of stud and channel shear connectors welded to the girders is given in Article 1.7.100. Channel shear connectors shall have at least $\frac{3}{16}$-in. fillet welds placed along the heel and toe of the channel.

The clear depth of concrete cover over the tops of the shear connectors shall be not less than 2 in. Shear connectors shall penetrate at least 2 in. above the bottom of slab.

The clear distance between the edge of a girder flange and the edge of the shear connectors shall be not less than 1 in.

1.7.98 *Effective Flange Width*

In composite girder construction, the assumed effective width of the slab as a T-beam flange shall not exceed the following:

1. One-fourth of the span length of the girder.
2. The distance center to center of girders.
3. Twelve times the least thickness of the slab.

For girders having a flange on one side only, the effective flange width shall not exceed one-twelfth of the span length of the girder, nor six times the thickness of the slab, nor one-half of the distance center to center of the next girder.

1.7.99 *Stresses*

Maximum compressive and tensile stresses in girders, which are not provided with temporary supports during the placing of the permanent dead load, shall be the sum of the stresses produced by the dead loads acting on the steel girders alone and the stresses produced by the superimposed loads acting on the composite girder. When girders are provided with effective intermediate supports, which are kept in place until the concrete has attained 75% of its required 28-day strength, the dead and live load stresses shall be computed on the basis of the composite section.

In continuous spans, the positive moment portion may be designed with composite sections as in simple spans. Shear connectors shall be provided in the negative moment portion in which the reinforcement steel embedded in the concrete is considered a part of the composite section. In case the reinforcement steel embedded in the concrete is not used in computing section properties for negative moments, shear connectors need not be provided in these portions of the spans, but additional connectors shall be placed in the region of the point of dead load contraflexure in accordance with Article 1.7.100(A)(3). Shear connectors shall be provided in accordance with Article 1.7.100.

1.7.100 *Shear*

A. HORIZONTAL SHEAR. The maximum pitch of shear connectors shall not exceed 24 in., except over the interior supports of continuous beams where wider spacing may be used to avoid placing connectors at locations of high stresses in the tension flange.

Resistance to horizontal shear shall be provided by mechanical shear connectors at the junction of the concrete slab and the steel girder. The shear connectors shall be mechanical devices placed transversely across the flange of the girder spaced at regular or variable intervals. The shear connectors shall be designed for fatigue and checked for ultimate strength.

1. Fatigue. The range of horizontal shear shall be computed by the formula

$$S_r = \frac{V_r Q}{I}$$

where S_r = the range of horizontal shear per linear inch at the junction of the slab and girder at the point in the span under consideration;

V_r = the range of shear due to live loads and impact. At any section, the range of shear shall be taken as the difference in the minimum and maximum shear envelopes (excluding dead loads);

Q = the statical moment about the neutral axis of the composite section of the transformed compressive concrete area or the area of reinforcement embedded in the concrete for negative moment;

I = the moment of inertia of the transformed composite girder in positive moment regions or the moment of inertia provided by the steel beam including or excluding the area of reinforcement embedded in the concrete in negative moment regions.

(In the above, the compressive concrete area is transformed into an equivalent area of steel by dividing the effective concrete flange width by the modular ratio, *n*.)

The allowable range of horizontal shear, Z_r, in pounds, on an individual connector is as follows:

CHANNELS $Z_r = Bw$

WELDED STUDS (for ratios of H/d equal to or greater than 4)

$$Z_r = \alpha d^2$$

where w = the length of a channel shear connector in inches measured in a transverse direction on the flange of a girder;
d = diameter of stud, in.;
α = 13,000 for 100,000 cycles
10,600 for 500,000 cycles
7,850 for 2,000,000 cycles;
B = 4,000 for 100,000 cycles
3,000 for 500,000 cycles
2,400 for 2,000,000 cycles;
H = height of stud, in.

The required pitch of shear connectors is determined by dividing the allowable range of horizontal shear of all connectors at one transverse girder cross section (ΣZ_r) by the horizontal range of shear S_r per linear inch. Over the interior supports of continuous beams, the pitch can be modified to avoid placing the connectors at locations of high stresses in the tension flange provided that the total number of connectors remains unchanged.

2. Ultimate Strength. The number of connectors so provided for fatigue shall be checked to ensure that adequate connectors are provided for ultimate strength. The number of shear connectors required between the points of maximum positive moment and the end supports or dead load points of contraflexure, and between points of maximum negative moment and the dead load points of contraflexure, shall equal or exceed the number given by the formula

$$N = \frac{P}{\phi S_u}$$

where N = the number of connectors between points of maximum positive moment and adjacent end supports or dead load points of contraflexure, or between points of maximum negative moment and adjacent dead load points of contraflexure;
S_u = the ultimate strength of the shear connector as given below;
ϕ = a reduction factor = 0.85;
P = force in the slab, defined hereafter as P_1, P_2, or P_3.

At points of maximum positive moment, the force in the slab is taken as the smaller value of the following formulas:

$$P_1 = A_s F_y$$

or

$$P_2 = 0.85 f'_c bc$$

where A_s = total area of the steel section including coverplates;
F_y = specified minimum yield point of the steel being used;
f'_c = compressive strength of concrete at 28 days;
b = effective flange width given in Article 1.7.99;
c = thickness of the concrete slab.

At points of maximum negative moment, the force in the slab is taken as

$$P_3 = A^r_s F^r_y$$

where A^r_s = total area of longitudinal reinforcing steel at the interior support within the effective flange width;
F^r_y = specified minimum yield point of the reinforcing steel.

The ultimate strength of the shear connector is given as follows:

CHANNELS $$S_u = 550\left(h + \frac{t}{2}\right) w \sqrt{f'_c}$$

WELDED STUDS ($H/d = 4$) $$S_u \geqq 930 d^2 \sqrt{f'_c}$$

where S_u = ultimate strength of individual shear connector, lb;
h = the average flange thickness of the channel flange, in.;
t = the thickness of the web of a channel, in.;
w = length of a channel shear connector, in.;
f'_c = compressive strength of the concrete at 28 days, psi;
d = diameter of stud, in.

3. Additional Connectors to Develop Slab Stress. The number of additional connectors required at points of contraflexure, when reinforcement steel embedded in the concrete is not used in computing section properties for negative moments, shall be computed by the formula

$$N_c = \frac{A_r f_r}{Z_r}$$

where N_c = number of additional connectors for each beam at point of contraflexure;
A_r = total area of longitudinal slab reinforcement steel for each beam over interior support;

f_r = range of stress due to live load plus impact in the slab reinforcement over the support (in lieu of more accurate computations, f_r can be taken as equal to 10,000 psi);

Z_r = the allowable range of horizontal shear on an individual shear connector.

The additional connectors, N_c, shall be placed adjacent to the point of dead load contraflexure within a distance equal to one-third of the effective slab width, that is, placed on either side of this point or centered about it.

B. VERTICAL SHEAR. The intensity of unit shearing stress in a composite girder can be determined on the basis that the web of the steel girder carries the total external shear, neglecting the effects of the steel flanges and of the concrete slab. The shear may be assumed to be uniformly distributed throughout the gross area of the web.

1.7.101 *Deflection*

The provisions of Article 1.7.12 in regard to deflections from live load plus impact also shall be applicable to composite girders.

When the girders are not provided with falsework or other effective intermediate support during the placing of the concrete slab, the deflection due to the weight of the slab and other permanent dead loads added before the concrete has attained 75% of its required 28-day strength shall be computed on the basis of noncomposite action.

Interim 8 Fatigue Stresses

Revision to Article 1.7.3

Delete the entire article and insert the following:

The number of cycles of maximum stress to be considered in the design shall be selected from Table 1.7.3A unless traffic and loadometer surveys or other considerations indicate otherwise.

Allowable fatigue stresses shall apply to those group loadings that include live load or wind load.

The number of cycles of stress to be considered for wind loads in combination with dead loads, except for structures where other considerations indicate a substantially different number of cycles, shall be 100,000 cycles.

Table 1.7.3A Stress Cycles: Main (Longitudinal) Load-Carrying Members

Type of Road	Case	ADTT[a]	Truck Loading	Lane Loading[b]
Freeways, expressways, major highways, and streets	I[c]	2,500 or more	over 2,000,000	500,000
	II	less than 2,500	500,000	100,000
Other highways and streets not included in Case I or II	III	—	100,000	100,000

Transverse Members and Details Subjected to Wheel Loads

Type of Road	Case	ADTT[a]	Truck Loading
Freeways, expressways, major highways, and streets	I[c]	2,500 or more	over 2,000,000
	II	less than 2,500	2,000,000
Other highways and streets	III	—	500,000

[a] Average daily truck traffic.
[b] Longitudinal members should also be checked for truck loading.
[c] This condition corresponds to an extremely heavily traveled artery.

Table 1.7.3B

Category (See Table 1.7.3C)	Allowable Range of Stress, F_{sr}, ksi			
	For 100,000 Cycles	For 500,000 Cycles	For 2,000,000 Cycles	For over 2,000,000 Cycles
A	60	36	24	24
B	45	27.5	18	16
C	32	19	13	10. 12[a]
D	27	16	10	7
E	21	12.5	8	5
F	15	12	9	8

[a] For transverse stiffener welds on girder webs or flanges.

Members and fasteners subject to repeated variations or reversals of stress shall be designed so that the maximum stress does not exceed the basic allowable stresses given in Articles 1.7.1 and 1.7.2, and that the actual range of stress does not exceed the allowable fatigue stress range given in Table 1.7.3B for the appropriate type and location of material in Table 1.7.3C and illustrated in Figure 1.7.3.

The range of stress is defined as the algebraic difference between the maximum stress and the minimum stress. Tension stress is considered to have the opposite algebraic sign from compression stress.

In Table 1.7.3C, "T" signifies range in tensile stress only, and "Rev." signifies a range of stress involving both tension and compression during a stress cycle.

Hybrid Girders

1.7.110 *Hybrid Girders—General Specifications*

This section pertains to the design of (a) noncomposite girders that have both flanges of the same minimum specified yield strength and a web with a lower minimum specified yield strength; (b) composite girders that have a tension flange with a higher minimum specified yield strength than the web and a compression flange with a minimum specified yield strength not less than that of the web; and (c) girders that utilize an orthotropic deck as the top flange and have a web with a lower minimum specified yield strength than the bottom flange. The design is applicable to both simple and continuous span girders. In noncomposite girders and in the negative moment portion of continuous span composite girders, the compression flange area shall be equal to the tension flange area or larger than the tension flange area by an amount not exceeding 15%. In composite girders, excluding the negative moment portion of continuous span girders, the compression flange area shall be equal to or smaller than the tension flange area. Steel girders that support the dead weight of the slab without composite action, but act compositely with the slab in supporting the live load, shall be considered to be composite girders. In either composite or noncomposite girders, the minimum specified yield strength of the web shall not be less than 35% of the minimum specified yield strength of the tension flange.

In girders that utilize an orthotropic deck as the top flange, the minimum specified yield strength of the web shall not be less than 35% of the minimum specified yield strength of the bottom flange in regions of positive bending moment, and not be less than 50% of the minimum specified yield strength of the bottom flange in regions of negative bending moment. As used in this section, flange refers to the flange of the steel girder and excludes the slab and reinforcing bars.

The provisions of Division I, Design, shall govern where applicable, except as specifically modified by Articles 1.7.110 through 1.7.113.

1.3.5 *Distribution of Loads and Design of Composite Wood-Concrete Members*

A. DISTRIBUTION OF CONCENTRATED LOADS FOR BENDING MOMENT AND SHEAR. For freely supported or continuous slab spans of composite wood-concrete construction, as described in Article 2.20.19(A), the wheel loads shall be distributed over a transverse width of 5 ft for bending moment and a width of 4 ft for shear.

For composite T beams of wood and concrete, as described in Article 2.20.19(B), the effective flange width shall not exceed that given in Article 1.7.98. Shear connectors shall be capable of resisting both vertical and horizontal movements.

B. DISTRIBUTION OF BENDING MOMENTS IN CONTINUOUS SPANS. Both positive and negative moments shall be distributed in accordance with the following table:

Maximum Bending Moments (% of Simple Span Moment)

	Maximum Uniform Dead Load Moments				Maximum Live Load Moments			
	Wood Subdeck		Composite Slab		Concentrated Load		Uniform Load	
Span	Pos.	Neg.	Pos.	Neg.	Pos.	Neg.	Pos.	Neg.
Interior	50	50	55	45	75	25	75	55
End	70	60	70	60	85	30	85	65
Two span[a]	65	70	60	75	85	30	80	75

[a] Continuous beam of two equal spans.

Impact should be considered in computing stresses for concrete and steel, but it should be neglected for wood.

C. DESIGN. The combination in a structural member of two elements having different mechanical properties requires the formulation of a design premise. Such a formulation as follows is based on the elastic properties of the materials:

E_c/E_w = 1 for slab in which the net concrete thickness is less than half the overall depth of the composite section;
E_c/E_w = 2 for slab in which the net concrete thickness is at least half the overall depth of the composite section;
E_s/E_w = 18.75 (for Douglas fir and Southern pine);

where E_c = modulus of elasticity of concrete;
E_w = modulus of elasticity of wood;
E_s = modulus of elasticity of steel.

Composite Concrete Flexural Members

1. Application

   Composite flexural members consist of concrete elements constructed in separate placements, but so interconnected that the elements respond to loads as a unit.

2. General Considerations

   a. The total depth of the composite member or portions thereof may be used in resisting the shear and the bending moment. The individual elements shall be investigated for all critical stages of loading.

   b. If the specified strength, unit weight, or other properties of the various components are different, the properties of the individual components, or the most critical values, shall be used in design.

   c. In calculating the flexural strength of a composite member by load factor design, no distinction shall be made between shored and unshored members.

   d. All elements shall be designed to support all loads introduced prior to the full development of the design strength of the composite member.

   e. Reinforcement shall be provided as necessary to prevent separation of the components.

3. Shoring

   When used, shoring shall not be removed until the supported

elements have developed the design properties required to support all loads and limit deflections and cracking at the time of shoring removal.

4. Vertical Shear
   a. When the total depth of the composite member is assumed to resist the vertical shear, the design shall be in accordance with the requirements of Article 1.5.29 or 1.5.35, as for a monolithically cast member of the same cross-sectional shape.
   b. Shear reinforcement shall be fully anchored in accordance with Article 1.5.21. Extended and anchored shear reinforcement may be included as ties for horizontal shear.
5. Horizontal Shear

   In a composite member, full transfer of the shear forces shall be assured at the interfaces of the separate components. Design for horizontal shear shall be in accordance with the requirements of Article 1.5.29(E) or 1.5.35(E).

J. T-Girder Construction

1. In T-girder construction, the girder web and slab shall be built integrally or otherwise effectively bonded together. Full transfer of shear forces shall be assured at the interface of web and slab. Where applicable, the design requirements of Article 1.5.23(I) for composite concrete members shall apply.
2. Compression Flange Width
   a. The effective slab width as a T-girder flange shall not exceed one-fourth of the span length of the girder, and its overhanging width on either side of the girder shall not exceed six times the thickness of the slab nor one-half of the clear distance to the next girder.
   b. For girders having a slab on one side only, the effective overhanging flange width shall not exceed one-twelfth of the span length of the girder, nor six times the thickness of the slab, nor one-half of the clear distance to the next girder.
   c. Isolated T girders in which the flange is used to provide additional compression area shall have a flange thickness not less than one-half of the width of the girder web and a total flange width not more than four times the width of the girder web.

3. Diaphragms

Diaphragms shall be placed between the girders at span ends, and within the spans at intervals not exceeding 40 ft. Diaphragms can be omitted where tests or structural analysis show adequate strength.

K. Box-Girder Construction

1. In box-girder construction, the girder web and top and bottom slab shall be built integrally or otherwise effectively bonded together. Full transfer of shear forces shall be assured at the interfaces of the girder web with the top and bottom slab. Design shall be in accordance with the requirements of Article 1.5.23(I). When required by design, changes in girder web thickness shall be tapered for a minimum distance of 12 times the difference in web thickness.
2. Compression Flange Width
   a. The effective slab width as a girder flange shall not exceed one-fourth of the span length of the girder, and its overhanging width on either side of the girder web shall not exceed six times the least thickness of the slab nor one-half of the clear distance to the next girder web.
   b. For girder webs having a slab on one side only, the effective overhanging flange width shall not exceed one-twelfth of the span length of the girder, nor six times the least thickness of the slab, nor one-half of the clear distance to the next girder web.
3. Top and Bottom Slab Thickness
   a. The thickness of the top slab shall be designed in accordance with Article 1.3.2(C), Case A, but shall be not less than 6 in.
   b. The thickness of the bottom slab shall not be less than one-sixteenth of the clear span between girder webs or $5\frac{1}{2}$ in., whichever is greater, except that the thickness need not be greater than the top slab unless required by design.
4. Top and Bottom Slab Reinforcement
   a. Minimum distributed reinforcement of 0.4% of the flange area shall be placed in the bottom slab parallel to the girder span. A single layer of reinforcement may be provided. The spacing of such reinforcement shall not exceed 18 in.

b. Minimum distributed reinforcement of 0.5% of the cross-sectional area of the slab, based on the least slab thickness, shall be placed in the bottom slab transverse to the girder span. Such reinforcement shall be distributed over both surfaces with a maximum spacing of 18 in. All transverse reinforcement in the bottom slab shall extend to the exterior face of the outside girder web in each group and be anchored by a standard 90° hook.

c. At least one-third of the bottom layer of the transverse reinforcement in the top slab shall extend to the exterior face of the outside girder web in each group and be anchored by a standard 90° hook. If the slab extends beyond the last girder web, such reinforcement shall extend into the slab overhang and shall have an anchorage beyond the exterior face of the girder web not less than that provided by a standard hook.

5. Diaphragms

Diaphragms or spreaders shall be placed between the girders at span ends, and within the spans at intervals not exceeding 60 ft. Diaphragms can be omitted where tests or structural analysis show adequate strength. Diaphragm spacing for curved girders shall be given special consideration.

1.5.24 *Design Methods*

A. The design of reinforced concrete members shall be made either with reference to service loads and allowable service load stresses as provided in "Service Load Design," or alternately, with reference to load factors and strengths as provided in "Load Factor Design."

B. All applicable provisions of this specification shall apply to both methods of design, except Articles 1.2.4 and 1.2.16 shall not apply for design by "Load Factor Design."

C. The strength and serviceability requirements of "Load Factor Design" can be assumed to be satisfied for design by "Service Load Design" if the service load stresses are limited to the values given in Article 1.5.26.

D. Members proportioned by "Load Factor Design" will sustain without damage at least the following overload:

Members designed for Group 1 loading $= D + 5/3(L + I)$
Members designed for Group 1A loading $= D + 2.2(L + I)$

1.5.26 *Allowable Service Load Stresses*

A. Concrete

For service load design, the stresses in concrete shall not exceed the following:

1. Flexure

   Extreme fiber stress in compression, f_c: $0.40f'_c$
   Extreme fiber stress in tension for plain concrete, f_t: $0.21f_r$
   Modulus of rupture, f_r, from tests, or if data are not available:
   Normal weight concrete: $7.5(f'_c)^{\frac{1}{2}}$
   "Sand-lightweight" concrete: $6.3(f'_c)^{\frac{1}{2}}$
   "All-lightweight" concrete: $5.5(f'_c)^{\frac{1}{2}}$
2. Shear

   Beams—shear carried by concrete, v_c: $0.95(f'_c)^{\frac{1}{2}}$
   Maximum shear carried by concrete plus shear reinforcement, v: $v_c + 4(f'_c)^{\frac{1}{2}}$
   Slabs and Footings (peripheral shear):
   Shear carried by concrete, v_c: $1.8(f'_c)^{\frac{1}{2}}$
   Maximum shear carried by concrete plus shear reinforcement, v: $3(f'_c)^{\frac{1}{2}}$
3. Bearing on loaded area, f_b: $0.30f'_c$
   a. When the supporting surface is wider on all sides than the loaded area, the allowable bearing stress on the loaded area can be increased by $(A_2/A_1)^{\frac{1}{2}}$, but not more than 2.
   b. When the supporting surface is sloped or stepped, A_2 may be taken as the area of the lower base of the largest frustrum of a right pyramid or cone contained wholly within the support and having for its upper base the loaded area, and having side slopes of 1 vertical to 2 horizontal.
   c. When the loaded area is subjected to high edge stresses due to deflection or eccentric loading, the allowable bearing stress on the loaded area shall be multiplied by a factor of 0.75. The requirements of (a) and (b) shall also apply.

B. Reinforcement

For service load design, the tensile stress in the reinforcement, f_s, shall not exceed the following:

Grade 40 or Grade 50 reinforcement: 20,000 psi
Grade 60 reinforcement: 24,000 psi

FATIGUE STRESS LIMIT. The range between a maximum and minimum stress in straight reinforcement caused by live load plus impact shall not exceed 21,000 psi. Bends in primary reinforcement shall be avoided in regions of high stress range.

1.5.27 *Flexure*

For investigation of service load stresses, the straight-line theory of stress and strain in flexure shall be used and the following assumptions shall be made:

a. A section plane before bending remains plane after bending; strains vary as the distance from the neutral axis.
b. The stress-strain relation of concrete is a straight line under service loads within the allowable service load stresses. Stresses vary as the distance from the neutral axis, except for deep flexural members with overall depth/span ratios greater than two-fifths for continuous spans and four-fifths for simple spans when a nonlinear distribution of stress should be considered.
c. The steel takes all the tension due to flexure.
d. The modular ratio, $n = E_s/E_c$, may be taken as the nearest whole number (but not less than 6). Except in calculations for deflections, the value of n for lightweight concrete shall be assumed to be the same as for normal weight concrete of the same strength.
e. In doubly reinforced flexural members, an effective modular ratio of $2E_s/E_c$ shall be used to transform the compression reinforcement for stress computations. The compressive stress in such reinforcement shall not be greater than the allowable tensile stress.

1.5.29 *Shear*

E. Horizontal Shear Design for Composite Concrete Flexural Members—Service Load

1. In a composite member, full transfer of the shear forces shall be assured at the interfaces of the separate components.
2. Full transfer of horizontal shear forces can be assumed when all of the following are satisfied: (a) the contact surfaces are clean and intentionally roughened, (b) minimum ties are provided in accordance with paragraph 6, (c) web members are designed to resist the entire vertical shear, and (d) all shear reinforcement is anchored into all intersecting components.

3. The horizontal shear stress, v_{dh}, can be computed at any cross section as

$$v_{dh} = \frac{V}{b_v d}$$

where d is the entire composite section. Alternatively, in any segment not exceeding one-tenth of the span, the actual change in compressive or tensile force to be transferred can be computed, and provisions can be made to transfer that force as horizontal shear to the supporting element.

4. The horizontal shear can be transferred at contact surfaces using the permissible horizontal shear stress, v_h, stated below.
   a. When ties are not provided, but the contact surfaces are clean and intentionally roughened, permissible $v_h = 36$ psi.
   b. When the minimum tie requirements of paragraph 6 are provided and the contact surfaces are clean but not intentionally roughened, permissible $v_h = 36$ psi.
   c. When the minimum tie requirements of paragraph 6 are provided and the contact surfaces are clean and intentionally roughened, permissible $v_h = 160$ psi.
   d. When v_{dh} exceeds 160 psi, design for horizontal shear shall be made in accordance with Article 1.5.29(D).
5. When tension exists perpendicular to any surface, shear transfer by contact can be assumed only when the minimum tie requirements of paragraph 6 are satisfied.
6. Ties for Horizontal Shear
   a. When vertical bars or extended stirrups are used to transfer horizontal shear, the tie area shall not be less than that required by Article 1.5.10(A)(2), and the spacing shall not exceed four times the least dimension of the supported element nor 24 in.
   b. Ties for horizontal shear may consist of single bars, multiple leg stirrups, or the vertical legs of welded wire fabric. All ties shall be adequately anchored into the components by embedment or hooks.
7. Measure of Roughness

   Internal roughness can be assumed only when the contact surface is roughened, clean, and free of laitance. Roughness shall have a full amplitude of approximately $\frac{1}{4}$ in.

4

Concrete-Concrete Composite Construction

SOURCE	Building Code Requirements for Reinforced Concrete (ACI318-71)
AVAILABLE FROM	American Concrete Institute P. O. Box 4754 Redford Station Detroit, Michigan 48219

Composite Concrete Flexural Members

17.0 *Notation*

b_v = the width of the cross section being investigated for horizontal shear
d = distance from extreme compression fiber to centroid of tension reinforcement, in.
v_{dh} = design horizontal shear stress at any cross section, psi
v_h = permissible horizontal shear stress, psi
V_u = total applied design shear force at section
ϕ = capacity reduction factor (see Section 9.2)

17.1 *Scope*

17.1.1 Composite concrete flexural members consist of concrete elements constructed in separate placements, but so interconnected that the elements respond to loads as a unit.

17.1.2 The provisions of all other chapters apply to composite concrete flexural members, except as specifically modified herein.

17.2 *General Considerations*

17.2.1 The entire composite member or portions thereof may be used in resisting the shear and the bending moment. The individual elements shall be investigated for all critical stages of loading.

17.2.2 If the specified strength, unit weight, or other properties of the various components are different, the properties of the individual components, or the most critical values, shall be used in design.

17.2.3 In calculating the strength of a composite member, no distinction shall be made between shored and unshored members.

17.2.4 All elements shall be designed to support all loads introduced prior to the full development of the design strength of the composite member.

17.2.5 Reinforcement shall be provided as necessary to control cracking and to prevent separation of the components.

17.2.6 Composite members shall meet the requirements for control of deflections given in Section 9.5.5.

17.3 *Shoring*

When used, shoring shall not be removed until the supported elements have developed the design properties required to support all loads and limit deflections and cracking at the time of shoring removal.

17.4 *Vertical Shear*

When the entire composite member is assumed to resist the vertical shear, the design shall be in accordance with the requirements of Chapter 11, as for a monolithically cast member of the same cross-sectional shape.

Web reinforcement shall be fully anchored into the components in accordance with Section 12.13. Extended and anchored web reinforcement may be included as ties for horizontal shear.

17.5 *Horizontal Shear*

17.5.1 In the composite member, full transfer of the shear forces shall be assured at the interfaces of the separate components.

17.5.2 Full transfer of horizontal shear forces can be assumed when all of the following are satisfied: (a) the contact surfaces are clean and intentionally roughened, (b) minimum ties are provided in accordance with Section 17.6.1, (c) web members are designed to resist the entire vertical shear, and (d) all stirrups are fully anchored into all intersecting components.

Otherwise, horizontal shear shall be fully investigated.

17.5.3 The horizontal shear stress, v_{dh}, can be calculated at any cross section as

$$v_{dh} = \frac{V_u}{\phi b_v d} \tag{17-1}$$

where d is for the entire composite section. Alternatively, the actual compressive or tensile force in any segment can be computed, and provisions can be made to transfer that force as horizontal shear to the supporting element. The ϕ factor specified for shear shall be used with the compressive or tensile force.

17.5.4 The design shear force can be transferred at contact surfaces using the permissible horizontal shear stresses, v_h, stated below.

a. When ties are not provided, but the contact surfaces are clean and intentionally roughened, permissible $v_h = 80$ psi.
b. When the minimum tie requirements of Section 17.6.1 are provided and the contact surfaces are clean but not intentionally roughened, permissible $v_h = 80$ psi.
c. When the minimum tie requirements of Section 17.6.1 are provided and the contact surfaces are clean and intentionally roughened, permissible $v_h = 350$ psi.
d. When v_{dh} exceeds 350 psi, design for horizontal shear shall be made in accordance with Section 11.15.

17.5.5 When tension exists perpendicular to any surface, shear transfer by contact can be assumed only when the minimum tie requirements of Section 17.6.1 are satisfied.

17.6 *Ties for Horizontal Shear*

17.6.1 When vertical bars or extended stirrups are used to transfer horizontal shear, the ties area shall not be less than that required by Section 11.1.2, and the spacing shall not exceed four times the least dimension of the supported element nor 24 in.

17.6.2 Ties for horizontal shear may consist of single bars, multiple leg stirrups, or the vertical legs of welded wire fabric. All ties shall be fully anchored into the components in accordance with Section 12.13.

17.7 *Measure of Roughness*

Intentional roughness can be assumed only when the interface is roughened with a full amplitude of approximately 1/4 in.

9.5 *Control of Deflections*

9.5.1 GENERAL Reinforced concrete members subject to bending shall be designed to have adequate stiffness to limit deflections or any deformations which may adversely affect the strength or serviceability of the structure at service loads.

9.5.2 NONPRESTRESSED ONE-WAY CONSTRUCTION

9.5.2.1 Minimum Thickness. The minimum thicknesses stipulated in Table 9.5(a) shall apply for one-way construction unless the computation of deflection indicates that lesser thickness may be used without adverse effects.

Table 9.5(a) Minimum Thickness of Beams or One-way Slabs unless Deflections Are Computed[a]

Member	Minimum Thickness, h			
	Simply Supported	One End Continuous	Both Ends Continuous	Cantilever
	Members not supporting or attached to partitions or other construction likely to be damaged by large deflections.			
Solid one-way slabs	$l/20$	$l/24$	$l/28$	$l/10$
Beams or ribbed one-way slabs	$l/16$	$l/18.5$	$l/21$	$l/8$

[a] The span length, l, is in inches.

The values given in this table shall be used directly for nonprestressed, reinforced concrete members made with normal weight concrete ($w = 145$ pcf) and Grade 60 reinforcement. For other conditions, the values shall be modified as follows:

1. For structural lightweight concrete having unit weights in the range 90–120 lb/ft^3, the values in the table shall be multiplied by $1.65–0.005w$, but not less than 1.09, where w is the unit weight in lb/ft^3.
2. For nonprestressed reinforcement having yield strengths other than 60,000 psi, the values in the table shall be multiplied by $0.4 + F_y/100{,}000$.

9.5.2.2 Computation of Immediate Deflection. Where deflections are to be computed, those which occur immediately on application of load shall be computed by the usual methods or formulas for elastic deflections. Unless values are obtained by a more comprehensive analysis, deflections shall be computed by taking the modulus of elasticity for concrete, as specified in Section 8.3.1 for normal weight or lightweight concrete, and by taking the effective moment of inertia as follows, but not greater than I_g:

$$I_e = \left(\frac{M_{cr}}{M_a}\right)^3 I_g + \left[1 - \left(\frac{M_{cr}}{M_a}\right)^3\right] I_{cr} \qquad (9\text{-}4)$$

where

$$M_{cr} = \frac{f_r I_g}{y_t} \qquad (9\text{-}5)$$

and

$$f_r = 7.5\sqrt{f'_c}$$

When lightweight aggregate concretes are used, one of the following modifications shall apply:

1. The equation for f_r shall be modified by substituting $f_{ct}/6.7$ for $\sqrt{f'_c}$, but the value of $f_{ct}/6.7$ used shall not exceed $\sqrt{f'_c}$. The value of f_{ct} shall be specified and the concrete proportioned in accordance with Section 4.2.
2. When f_{ct} is not specified, the equation for f_r shall be multiplied by 0.75 for "all-lightweight" concrete, and 0.85 for "sand-lightweight" concrete. Linear interpolation can be used when partial sand replacement is used.

For continuous spans, the effective moment of inertia can be taken as the average of the values obtained from Equation 9-4 for the critical positive and negative moment sections.

9.5.2.3 Computation of Long-Time Deflection. Unless values are obtained by a more comprehensive analysis, the additional long-time deflection for both normal weight and lightweight concrete flexural members shall be obtained by multiplying the immediate deflection caused by the sustained load considered, computed in accordance with Section 9.5.2.2, by the factor

$$2 - 1.2\left(\frac{A'_s}{A_s}\right) \geq 0.6$$

Table 9.5(b) Maximum Allowable Computed Deflections

Type of Member	Deflection To Be Considered	Deflection Limitation
Flat roofs not supporting or attached to nonstructural elements likely to be damaged by large deflections	Immediate deflection due to the live load, L	$\frac{l}{180}$[a]
Floors not supporting or attached to nonstructural elements likely to be damaged by large deflections	Immediate deflection due to the live load, L	$\frac{l}{360}$
Roof or floor construction supporting or attached to nonstructural elements not likely to be damaged by large deflections	That part of the total deflection which occurs after attachment of the nonstructural elements, the sum of the long-time deflection due to all sustained loads, and the immediate deflection due to any additional live load[b]	$\frac{l}{480}$[c]
Roof or floor construction supporting or attached to nonstructural elements likely to be damaged by large deflections		$\frac{l}{240}$[d]

[a] This limit is not intended to safeguard against ponding. Ponding should be checked by suitable calculations of deflection, including the added deflections due to ponded water, and considering long-time effects of all sustained loads, camber, construction tolerances, and reliability of provisions for drainage.

[b] The long-time deflection shall be determined in accordance with Section 9.5.2.3 or 9.5.4.2, but it may be reduced by the amount of deflection which occurs before attachment of the nonstructural elements. This amount shall be determined on the basis of accepted engineering data relating to the time-deflection characteristics of members similar to those being considered.

[c] This limit may be exceeded if adequate measures are taken to prevent damage to supported or attached elements.

[d] This limit may not be greater than the tolerance provided for the nonstructural elements. This limit may be exceeded if camber is provided so that the total deflection minus the camber does not exceed the limitation.

9.5.2.4 Allowable Deflection. The deflection computed in accordance with Sections 9.5.2.2 and 9.5.2.3 shall not exceed the limits stipulated in Table 9.5(b).

9.5.5 COMPOSITE MEMBERS

9.5.5.1 Shored Construction. If composite members are supported during construction in such a manner that, after removal of temporary supports, the dead load is resisted by the full composite section, the composite member may be considered equivalent to a cast-in-place member for the purposes of deflection calculation. For nonprestressed members, the portion of the member in compression shall determine whether the values given in Table 9.5(a) for normal weight or lightweight concrete shall apply. If deflection is calculated, account should be taken of the curvatures resulting from differential shrinkage of the precast and cast-in-place components, and of the axial creep effects in a prestressed concrete member.

9.5.5.2 Unshored Construction. If the thickness of a nonprestressed, precast member meets the requirements of Table 9.5(a), deflection need not be computed. If the thickness of a nonprestressed composite member meets the requirements of Table 9.5(a), deflection occurring after the member becomes composite need not be calculated, but the long-time deflection of the precast member should be investigated for the magnitude and duration of load prior to the beginning of effective composite action.

9.5.5.3 Allowable Deflection. The deflection computed in accordance with the requirements of Sections 9.5.5.1 and 9.5.5.2 shall not exceed the limits stipulated in Tablc 9.5(b).

10.10 *Composite Compression Members*

10.15.1 Composite compression members shall include all concrete compression members reinforced longitudinally with structural steel shape, pipe, or tubing with or without longitudinal bars.

10.15.2 The strength of composite compression members shall be computed for the same limiting conditions applicable to ordinary reinforced concrete members. Any direct compression load capacity assigned to the concrete in a member must be transferred to the concrete by members or brackets in direct bearing on the compression member concrete. All compression load capacity not assigned to the concrete shall be developed by direct connection to the structural steel shape, pipe, or tube.

10.15.3 For slenderness calculations, the radius of gyration of the composite section shall not be greater than the value given by

$$r = \sqrt{\frac{1/5(E_cI_g + E_sI_t)}{1/5(E_cA_g + E_sA_t)}} \tag{10-10}$$

For computing P_c in Equation 10-6, EI of the composite section shall not be greater than

$$EI = \sqrt{\frac{E_cI_g/5 + E_sI_t}{I + \beta_d}} \tag{10-11}$$

10.15.4 Where the composite compression member consists of a steel-encased concrete core, the thickness of the steel encasement shall be greater than

$$b\sqrt{\frac{f_y}{3E_s}}, \quad \text{for each face of width } b \tag{10-12}$$

$$h\sqrt{\frac{f_y}{8E_s}}, \quad \text{for circular sections of diameter } h \tag{10-13}$$

Longitudinal bars within the encasement may be considered in computing A_t and I_t.

10.15.5 Where the composite compression member consists of a spiral-bound concrete encasement around a structural steel core, f'_c shall not be less than 2500 psi and spiral reinforcement shall conform to Equation 10-3. The design yield strength of the structural steel core shall be the specified minimum yield strength for the grade of structural steel used, but not to exceed 50,000 psi. Longitudinal reinforcing bars within the spiral shall not be less than 0.01 nor more than 0.08 times the net concrete section and may be considered in computing A_t and I_t.

10.15.6 Where the composite compression member consists of laterally tied concrete around a structural steel core, f'_c shall not be less than 2500 psi, and the design yield strength of the structural steel core shall be the specified minimum yield strength for the grade of structural steel used, but not to exceed 50,000 psi. Lateral ties shall extend completely around the steel core. Lateral ties shall be No. 5 bars, or smaller bars having a diameter not less than one-fiftieth of the longest side or diameter of the cross section, but not smaller than No. 3 bars. The vertical spacing of lateral ties shall not exceed one-half of the least width of the cross section, or 48 tie bar diameters, or 16 longitudinal bar diameters. Welded wire fabric of equivalent area can be used.

Longitudinal reinforcing bars within the ties, not less than 0.01 nor more than 0.08 times the net concrete section, shall be provided. These shall be spaced not greater than one-half of the least width of the cross section. A longitudinal bar shall be placed at each corner of a rectangular cross section. Bars placed within the lateral ties can be considered in computing A_t for strength calculations, but not I_t for slenderness calculations.

Composite Concrete Flexural Members Commentary

17.1 *Scope*

17.1.1 The scope of this chapter is intended to include all types of composite concrete flexural members, including composite single-T or double-T members, box sections, folded plates, lift slabs, and other structural elements, all of which should conform to the provisions of this chapter. In some cases, with fully cast-in-place concrete, it may be necessary to design the interface of consecutive placements of concrete as required for composite members. Composite structural steel-concrete members are not covered in this chapter, since such sections are covered in the *Specification for the Design, Fabrication and Erection of Structural Steel for Buildings*, published by the American Institute of Steel Construction (AISC).

17.1.2 The code in its entirety applies to composite concrete flexural members, except as specifically modified in Chapter 17. For instance, deep composite sections shall be designed in accord with Sections 10.7 and 11.9. When composite concrete flexural members are subjected to axial loads, Sections 10.8, 10.9, and 10.10 (or 10.11) apply. The alternate design method of Section 8.10 can also be used.

17.2 *General Considerations*

17.2.1 This section permits the designer to use any or all of the various components in supporting the load in the most expeditious manner.

17.2.3 Tests to destruction indicate no difference in strength of shored and unshored members.

17.2.5 The extent of cracking permitted is dependent on such factors as environment, aesthetics, and occupancy. In addition, composite action must not be impaired.

17.2.6 The premature loading of precast elements can cause excessive deflections as the result of creep and shrinkage. This is especially so at early ages when the moisture content is high and the strength is low.

The transfer of shear by direct bond is essential if excessive deflection from slippage is to be prevented. A shear key is an added mechanical factor of safety, but it cannot operate until slippage occurs.

17.3 *Shoring*

The provisions of Sections 9.5.5.1 and 9.5.5.2 must be considered with regard to deflections of shored and unshored members. Before shoring is removed, it should be ascertained that the strength and serviceability characteristics of the structure will not be impaired.

17.5 *Horizontal Shear*

17.5.1 The full transfer of horizontal shear between segments must be ensured by contact stresses or properly anchored ties, or both.

17.5.2 Tests (17.1) indicate that horizontal shear does not present a problem in T beams when the portion below the flange is designed to resist the vertical shear, the interfaces of the components are rough, and minimum ties are provided according to Section 17.6.1. The ties must be extended across the joint and fully anchored on both sides of the joint in accord with Section 12.13. These considerations may be used with other segmental shapes.

17.5.3 The calculated horizontal shear stress represents the force per unit area of interface. When the design is developed using the alternate method of Section 8.10, V_u is the shear due to dead and live loads calculated using unity load and ϕ factors. Also, when this method is used and a combination of gravity, wind, and earthquake loads govern, Section 8.10.5 applies.

17.5.4 The permissible horizontal shear stresses, v_h, apply when the design is based on the load and ϕ factors of Chapter 9. When the alternate design method of Section 8.10 is used, the value of v_h should be reduced in accordance with the provisions for shear stresses in Section 8.10.3.

In reviewing composite concrete flexural members for serviceability at service loads and for handling at construction loads, V_u may be replaced by the service load shear or handling load shear in Equation 17-1. The resulting service load or handling load horizontal shear stress should be compared with the allowable stresses ($0.55v_h$) to insure that an adequate factor of safety results.

17.5.5 Proper anchorage of bars extending across joints is required to insure that contact of the interfaces is maintained.

17.6 *Ties for Horizontal Shear*

The minimum areas and maximum spacings are based on test data given in Reference 17.1.

17.7 *Measure of Roughness*

This section conforms with the provisions of Chapter 11, and is based on tests discussed in Reference 17.1.

9.5.5 *Composite Members*

Since few tests have been made to study the immediate and long-time deflections of composite members, the rules given in Sections 9.5.5.1 and 9.5.5.2 are based on the judgment of the committee and on experience.

If any portion of a composite member is prestressed or if the member is prestressed after the components have been cast, the provisions of Section 9.5.4 apply and deflections shall always be calculated. For nonprestressed members, deflections need to be calculated and compared with the limiting values in Table 9.5(b) only when the thickness of the member or the precast part of the member is less than the minimum thickness given in Table 9.5(a). In unshored construction, the thickness of concern depends on whether the deflection before or after the attainment of effective composite action is being considered.*

Table 9.5(a), based on the use of reinforcement having yield strengths of 60,000 psi, is referred to only in Sections 9.5.2 and 9.5.5. It can be used in lieu of calculation of deflections only for the types of members covered by those sections and only if these members do *not* support and are *not* attached to partitions or other construction likely to be damaged by large deflections.

The notes beneath the table are essential to its use for reinforced concrete members constructed with structural lightweight concrete and/or with reinforcement having a yield strength not equal to 60,000 psi. If both of these conditions exist, the corrections in footnotes *a* and *b* shall both be made.

The modification for lightweight concrete in footnote *a* is based on studies of the results and discussions in Reference 9.5. No correction is given for concretes weighing between 120 and 145 lb/ft^3 since the correction term would be close to unity in this range.

*In Chapter 17, it is stated that distinction need not be made between shored and unshored members. This refers to strength calculations, not to deflections.

The modifications for yield strength in footnote *b* is based on judgment, experience, and studies of the results of tests and of unpublished analyses. The simple expression given is approximate, but it should yield conservative results for the types of members considered in the table, for typical reinforcement ratios, and for values of F_y between 40 and 80 ksi.

If the minimum thickness obtained using this table is considered excessive, the designer has the option of computing deflections in accordance with Sections 9.5.1 and 9.5.5.

Table 9.5(b) is the result of an effort by Committee 318 to simplify the very extensive set of limitations which would be required to cover all possible types of construction and conditions of loading. It should be noted that for members supporting or attached to other elements, the limitations given in this table relate only to supported or attached *nonstructural* elements. For those structures in which structural members are likely to be affected by deflection or deformation of members, to which they are attached in such a manner as to affect adversely the strength of the structure, these deflections and the resulting forces should be considered explicitly in the analysis and design of the structures as required by Section 9.5.1.

When long-time deflections are computed, the portion of the deflection before attachment of the nonstructural elements can be deducted. In making this correction, use can be made of the curves in Figure 9-2 for members of usual sizes and shapes.

Index